Plant Science Research and Practices

Plant Dormancy

Mechanisms, Causes and Effects

PLANT SCIENCE RESEARCH AND PRACTICES

Additional books and e-books in this series can be found on Nova's website under the Series tab.

PLANT SCIENCE RESEARCH AND PRACTICES

PLANT DORMANCY

MECHANISMS, CAUSES AND EFFECTS

RENATO V. BOTELHO
EDITOR

science publishers
New York

Library of Congress Cataloging-in-Publication Data

Names: Botelho, Renato V. (Renato Vasconcelos), editor.
Title: Plant dormancy : mechanisms, causes and effects / editor: Renato V. Botelho.
Description: Hauppauge, New York : Nova Science Publishers, Inc., 2019. | Series: Plant science research and practices | Includes bibliographical references and index.
Identifiers: LCCN 2019017076 (print) | LCCN 2019018798 (ebook) | ISBN 9781536153811 | ISBN 9781536153804 (hardcover : alk. paper)
Subjects: LCSH: Dormancy in plants. | Fruit--Dormancy.
Classification: LCC QK761 (ebook) | LCC QK761 .P5788 (print) | DDC 571.7/82--dc23
LC record available at https://lccn.loc.gov/2019017076

Published by Nova Science Publishers, Inc. † New York

CONTENTS

PREFACE

Dormancy is a mechanism found in several plant species developed through evolution, which allows plants to survive in adverse conditions and ensure their perpetuation. This mechanism, however, can represent a barrier that can compromise the development of the species of interest, and therefore, the success of its cultivation. It is important to understand the causes, effects, and mechanisms, as well as the technologies for overcoming dormancy of buds, seeds, and plant propagules. In this context, this publication has as a main objective of offering an in-depth view of the dormancy process in cultivated plants, based on a survey of more recent international research on the subject.

Written by several researchers from different institutions of education and research, this book presents information on the most diverse species of plants of agronomic and forestry interest. In addition to the physiological, biochemical, and genetic aspects, the chapters also address the possible causes of dormancy, as well as technological issues that are essential for the successful cultivation of several economically important species.

The focus of this publication is on temperate fruit species. Classical examples of temperate plants that present dormancy are fruit trees such as apple, pear, quince, apricot, and cherry. These pome and stone fruits are typical representatives of a deciduous temperate climate, wherein trees drop their leaves at the end of the reproductive cycle, consequently entering dormancy.

Dormancy has its control effected by factors of a hormonal nature, where the hormone promoters and growth inhibitors are balanced. Edaphoclimatic conditions also affect plant dormancy periods, for the most part through temperature changes. Other chapters in this book are dedicated especially to potatoes, vegetables, as well as forest and forage species, which makes this text a complete and comprehensive work on the subject.

Chapter 1 - Dormancy is a multiple and complex phenomenon which multiplicity stems from the fact that it affects several plant structures. Surviving an unfavourable season - such as the winter time of the year, mainly due to low temperatures and even a possible thick snow cover - implies in many manifestations that interfere in dormancy: formation of protection structures, accumulation of metabolites in specific tissues, mechanisms of resistance, among others. Classical examples of temperate plants that present dormancy are fruit species such as apple, pear, quince, apricot and cherry trees. These pome and stone fruits are typical representatives of deciduous temperate climate, which drop their leaves at the end of the reproductive cycle, consequently entering in dormancy. Dormancy has its control effected by factors of hormonal nature from a regulation where the hormones promoters and the growth inhibitors are balanced. Edaphoclimatic conditions, for the most part by the temperature, also affect the plants marking the entrance and the exit of the dormancy period. Types or classes of dormancy as ecodormancy, endodormancy and paradormancy, may interact with each other chronologically, and when it happens, they therefore constitute phases within the total dormancy period, in addition to involving some common control mechanisms. The multiplicity of facts and phenomena that occur in the plant in the pre-dormancy period until its overcoming, subject to hormonal balances and imbalances, climatic and soil conditions, and added to anthropic intervention activities, such as cultural practices, are responsible for a new life cycle - growth and reproduction - of deciduous plants of temperate climate.

Chapter 2 - The use of methods to estimate chilling requirement of temperate fruit trees are important tools for the prediction of physiological phase changes in the plant, as well as the decision making on the cultural management to be carried out in the orchards. Several methods can be used to estimate chilling requirement of temperate fruit trees. They can be basically divided into biological methods and temperature modeling methods. Biological methods are those characterized by direct phenological observation (sprouting and flowering) of plant material (parts of plants or whole plants) submitted to controlled and distinct conditions during dormancy. In general, this method is used in experiments to characterize genotypes with respect to the chilling requirement or for studies of temperature modeling, evaluating the behavior of the buds in different situations. On the other hand, the methods of temperature modeling have the objective of measuring the amount of chilling accumulation during the winter period and, from the characterization of the need of the different cultivars, it is possible to predict the time of budbreak. Although widely used and an important tool in the management of fruit trees, this second method presents a very variable performance in certain conditions and crops, since the methodologies developed, for the most part, were based on empirical observations and not on physiological processes. Advances in studies on the mechanisms involved in tree dormancy control contribute to the development of more accurate models, increasing efficiency in predicting the physiological development of trees and improving information for decision-making in orchards management.

Chapter 3 - Bud dormancy of temperate species is a natural and necessary cyclic process that occurs as part of a plant protection system against low temperatures during the winter. Several climatic factors are involved in this process, with emphasis on temperature, and internal factors such as phytohormones, carbohydrates, growth inhibitors, water content, enzymes and oxidative stress. Due to the interaction among these factors, the study of dormancy is complex and several methods have been developed to evaluate the presence and intensity of dormancy. In this chapter some of these methods will be described, discussing how they can be performed and their results. The Tabuenca's test is used for evaluation of floral buds dormancy. The single node cutting test is one of the most used to evaluate dormancy intensity. The dormancy index was created to more clearly characterize the dormancy intensity from the results of the single node cutting test. Biochemical tests can be performed to quantify various substances involved in the dormancy process, such as carbohydrates, enzymes and reactive oxygen species. The tetrazolium test is a newly developed test to rapidly quantify the respiratory activity of buds.

Chapter 4 - The best and most efficient way to avoid problems related to dormancy on deciduous fruit trees is the use of cultivars adapted to the amount of winter cold in the growing region. In addition to the use of cultivars with low chilling requirement, there are some technical interventions that can minimize the effects of lack of chilling on low sprouing and flowering, such as artificial chilling, ring incision, arching of branches, growth control, post-harvest pruning and defoliation. After the nursery trees are removed from the soil, they can be stored in a cold room for a period of 30 to 45 days at 2 to 6 °C, allowing a uniform sprouting and greater branches growth. The ringing can be used, in the first and second year, for the plant in formation of the upper layers when a suitable sprouting for the plant formation did not occur. In the climatic conditions of southern Brazil, the period of vegetative development of temperate fruit trees is superior to that observed in typical regions of temperate climate, which associated with high temperatures and high rainfall rates during the cycle can result in excessive branch growth. Under such conditions, the use of management techniques that restrict vegetative development becomes essential. Among the measures to control the vegetative growth of the branches of the year, the use of calcium prohexadione is efficient in apple, pear and plum tree crops. In mild climatic conditions the temperate fruit trees resume the vegetative growth after harvesting, under these conditions pruning when the plants still have leaves reduces the plant vigor and photoassimilates are reallocated to the reproductive organs that will remain in the plant. Arching provides a weakening of the branches, favoring sprouting. Branches ringed to interrupt the normal circulation of sap also favor sprouting. In regions of subtropical climate or in years that the temperatures are not too low in the autumn-winter, the natural fall of leaves does not occur. Under these conditions it is necessary to make chemical defoliation, which should preferably be carried out with chemicals that promote leaf senescence.

Chapter 5 - Hibernal dormancy of vines are characterized by a plant state of physiological inactivity, which allows its survival under adverse conditions. In this period, the plant does not show vegetative growth, but its metabolic activities continues, although reduced. For these vines to resume the growth naturally, its exposure to a certain period of low temperatures is required, called chilling units. The dormancy presents progressive development that begins during the autumn, increasing its intensity until reaching the deep dormancy or endodormancy, conferring the capacity of survival in temperatures unsuitable to vegetal development. Thus, one of the major limitations for grapevine cultivation in tropical and subtropical regions, where the amount of hibernal chilling is not enough, is the overcoming of your period of dormancy. In these regions, the application of some substances is a strategy that can be used to induce bud burst, flowering and ripening of temperate climate fruit species like vine. Among several substances that have an effect on artificial induction of bud burst, the authors highlight the hydrogen cyanamide, successfully used in several vine growing regions, as well as other promising products such as mineral and vegetable oils and garlic extract. This chapter approaches the main aspects related to grapevine dormancy, as well as the strategies for induction of a new cycle in tropical and subtropical regions through application of bud burst inducers.

Chapter 6 - In a world with increasing population, food production to feed this population growth is crucial. Potato ranks third as the main source of food in the world. In potato's production in the field, the use of tuber seed with appropriated sprout growth is important to achieve high yields. The start of the sprout growth of potato tubers, in turn, is dependent on the breaking of dormancy. In addition, after tuber harvest, it is important to maintain tuber dormancy in order to keep the quality of this tuber. In this chapter, the authors stated the definition of potato tuber dormancy, showed the influence of dormancy on tuber yield, gave the main managements to keep or break tuber dormancy and reported on the factors affecting dormancy period. In the sequence the authors discuss the physiological aspects of tuber dormancy, i.e., the main plant hormones that are related to the regulation of this process. Finally, the authors present the genetics of tuber dormancy and release, presenting the newest findings in this area of study and the future steps to achieve a better understanding of the genetic control of the breaking of dormancy in potato tuber. The authors hope that this text would help technicians and growers to better understand and manage this important food crop.

Chapter 7 - Of the US $ 4.6 billion in world trade of vegetable seeds, Europe and Asia account for 70%. Of this amount, Brazil is responsible for moving about US $ 100 million. Considered by several experts as an excellent business environment in the world context of the seed industry, Brazil manages to combine its agricultural vocation with a good technological system. Furthermore, the Country has a continental characteristic, providing suitable climatic conditions for the production of high quality seeds of tropical and temperate species. Among the difficulties encountered in the seed production sector, dormancy stands out, which is usually an undesirable characteristic in the olericulture, since the rapid germination and growth are required as a primordial factor. The knowledge of dormancy mechanisms and their duration in the different vegetable seeds are ecologically and economically important, since they help in the selection of specific treatments that act on the metabolism of the seed, releasing the embryo for development or making it apt to germinate. Dormancy allows distribution of germination time, and is also an adaptive feature that ensures the survival and perpetuation of the species in the different ecosystems. The germinative delay maintains the species at the seed stage, i.e., at the stage of the plant cycle in which its resistance to the unfavorable conditions is greater. Seed dormancy is controlled by the genotype associated with: environmental factors, such as temperature; physical factors, such as tegument thickness; and regulation of hormones such as abscisic acid and gibberellin. Generally, plants produce seeds with various levels of dormancy. This aspect helps to distribute their descendants through time and space. In addition, it allows the seeds not to germinate during the adverse periods, guaranteeing a time required for the dispersion to the greater geographic distance, providing the survival of the species. In order to standardize and potentiate the germination of some seeds, it is necessary to identify the type of dormancy. Thus, this chapter presents several specific methods for breaking dormancy in vegetable seeds, considering mainly aspects of physiological and morpho-anatomic of the seeds.

Chapter 8 - The Mixed Ombrophilous Forest (MOF), also known as Araucaria Forest, is a phytophysiognomy located within the Atlantic Forest, the fourth most important biodiversity hotspot on the planet. Due to the great diversity of species present in this phytophysiognomy, it is important to know its ecological characteristics, as is the case of seed dormancy. In the MOF, most of the seeds are classified as orthodox, present physical dormancy, low to moderate germination. Several techniques are used to overcome dormancy of forest seeds. The procedure is performed to accelerate germination process, among techniques, chemical and mechanical scarification are the most applied in seeds of the MOF species, due to high percentage of seeds with physical dormancy. It is verified that the majority of species that present dormancy belongs to the families Fabaceae and Lauraceae, mainly the species of the genus *Ocotea*, that are present in the forest upper stratum. In addition, some seeds require pre-germination treatment, even if they do not present dormancy. Therefore, the information on seed dormancy of the species of the MOF generating an important tool for better decision making in management of forest recovery techniques.

Chapter 9 - In forage legumes, seeds with physical dormancy (hard seeds) are not exceptional and during the germination test their embryos do not imbibe or germinate due to a physical impediment of their seed coats. This paper presents aspects related to physical dormancy studied in alfalfa (*Medicago sativa* L.) and white clover (*Trifolium repens* L.), and the results are compared, showing similarities and differences between the samples of both species. The studies carried out in alfalfa seeds of cv. Baralfa 85 and white clover seeds of cv. NK Churrinche (2004 and 2006 harvests) analyzed the following aspects: weight and moisture content, passage of water to the embryo, viability, anatomical characteristics, determinations of structural substances, polyphenols, tannins, cutin and ions present in the testa of seeds with and without physical dormancy. The comparative analysis indicates that both species have physical dormancy with different levels of depth, which is overcome over time. The amount of seeds that break dormancy during the period when they are immersed in water is different between species. Physical dormancy is related to seed moisture content and modifications that occur in the seed coat, which differ between seeds without dormancy and hard seeds. The different degrees of depth of physical dormancy and the differences between species can be explained in terms of the characteristics of the testa: thicker cuticle and longer macrosclereids, thicker cell wall, and more conspicuous osteosclereids and in higher numbers, as well as accumulation of lignins, tannins, polyphenols and pectic substances, and the amount of cellulose and hemicellulose. The ions found can be related to the components of the cell wall in which they are found and the cementing properties of these substances. Physical dormancy is a mechanism that protects the embryo and, together with the low moisture content found in hard seeds, ensures longevity, prevents deterioration and predisposes most of the embryos to develop into normal seedlings. These characteristics of the seed coat prevent dehydration of the embryo and slow down exchanges with relative humidity of the environment, favoring good seed storage. A fraction of white clover seeds presented deeper dormancy than in alfalfa.

Chapter 10 - Dormancy is a resistance or adaptive mechanism that plants have developed as a way of surviving adversities during the winter period. Dormancy is not a rapidly acquired mechanism, but a progressive development process that begins during the fall, increasing in intensity until reaching the so-called deep dormancy or endodormancy during the winter. This phase can be induced by internal and external factors and is regulated by their interactions: environmental factors, such as photoperiod and temperature; and by physiological factors that include phytohormones and biochemicals (sugar, starch and carbohydrate content). The intensity and the way this phenomenon manifests in temperate plants, different growing environments, has an important impact on the control, maintenance and production of such species. Thus, a better understanding of the types of dormancy (paradormancy, endodormancy and ecodormancy) and the characterization of the growing environment are important to obtain satisfactory yields and to determine the most suitable agronomic techniques. This chapter presents to the reader the internal and external aspects related to the induction and exit of dormancy, as well as methods of evaluating the stage of dormancy.

In: Plant Dormancy
Editor: Renato V. Botelho
ISBN: 978-1-53615-380-4

Chapter 1

EVOLUTION OF DORMANCY: PROCESSES OF INDUCTION AND OVERCOMING IN FRUIT TREES OF TEMPERATE CLIMATE

Leo Rufato[1,*] ***and Augusto José Posser***[1]
[1]Agroveterinary Sciences Center, Santa Catarina State University, Lages, Santa Catarina State, Brazil

ABSTRACT

Dormancy is a multiple and complex phenomenon which multiplicity stems from the fact that it affects several plant structures. Surviving an unfavourable season - such as the winter time of the year, mainly due to low temperatures and even a possible thick snow cover - implies in many manifestations that interfere in dormancy: formation of protection structures, accumulation of metabolites in specific tissues, mechanisms of resistance, among others. Classical examples of temperate plants that present dormancy are fruit species such as apple, pear, quince, apricot and cherry trees. These pome and stone fruits are typical representatives of deciduous temperate climate, which drop their leaves at the end of the reproductive cycle, consequently entering in dormancy. Dormancy has its

[*] Corresponding Author's E-mail: leoruffato@yahoo.com.br.

control effected by factors of hormonal nature from a regulation where the hormones promoters and the growth inhibitors are balanced. Edaphoclimatic conditions, for the most part by the temperature, also affect the plants marking the entrance and the exit of the dormancy period. Types or classes of dormancy as ecodormancy, endodormancy and paradormancy, may interact with each other chronologically, and when it happens, they therefore constitute phases within the total dormancy period, in addition to involving some common control mechanisms. The multiplicity of facts and phenomena that occur in the plant in the pre-dormancy period until its overcoming, subject to hormonal balances and imbalances, climatic and soil conditions, and added to anthropic intervention activities, such as cultural practices, are responsible for a new life cycle - growth and reproduction - of deciduous plants of temperate climate.

Keywords: decidous fruit trees, plant regulators, cultural practices, endodormancy

1. Introduction

Dormancy is a multiple and complex phenomenon. Its multiplicity stems from the fact that it affects several plant structures, which appeared at different times during evolution and evolved in a variety of biotypes. In addition, the complexity also results from the numerous anatomical and physiological relationships of the properly dormant structure with its adjacent parts in the plant (Viémont 2000).

Millions of years ago, dormancy had a morphogenic function of spatial and temporal organization of plants (Viémont 2000) before becoming, as it now stands, a mechanism of survival against a possibly unfavourable season. However, this latter aspect has become a major concern related to plant physiology regarding the control of sprouting, growth and reproduction of temperate species (Viémont 2000).

Plants are not able to withstand low temperatures while they are growing and developing, nor do they have the capacity to acquire this resistance when they are already in conditions of low temperatures. In cold and altitude zones, plants developed this period of dormancy to survive, mainly, to the cold, and thus they evolved in the time. Therefore, in their natural habitat, plants are rarely damaged or affected by cold.

Deciduous fruit species are naturally adapted to places with well-defined climatic seasons with spring and summer periods that provide them with adequate temperatures for vegetative growth, as well as low-temperature in autumn and winter periods which, in addition to not allowing visible vegetative growth, could destroy more tender tissues (Salaya 2009).

Dormancy has a famous commercial importance, which has intensified since the 1950s. Growers in temperate regions coexist with the risk of a possible premature overcoming of dormancy in orchards, which would expose plants to the risk of late frost or even spring frosts, which damage active meristems, flowers and, in some cases, small fruits (Sozzi et al. 2007).

The cultivation and management techniques have become, in this case, an important factor to increase the natural capacity of the species to survive the cold seasons. Management of irrigation and fertilization to avoid late growth are examples of damage prevention, since many cultivated species were breeded to obtain specific factors such as fruit quality and selected in their natural habitats and then introduced in climates different from those in which they had evolved (Westwood 1982).

Surviving an unfavourable season - such as the winter time of the year, mainly due to low temperatures and even a possible thick snow cover - implies in many manifestations that interfere in dormancy, such as formation of protection structures, accumulation of metabolites in specific tissues and mechanisms of resistance (Viémont 2000).

Numerous studies are currently underway on the inducing and overcoming aspects of plant dormancy, such as the dormancy site in the plant, photoperiod and induction of dormancy by environmental conditions; physiology of dormancy control, specific phytormones; cold temperatures, effective temperatures, differentiation of buds, modification of the need for cold hours by the environment and/or cultural practices, models to calculate the need for chilling hours, chemical products to break dormancy, and so forth.

Classical examples of plants that present dormancy are fruit trees such as apple, pear, quince, apricot and cherry trees. These pome and stone fruits are typical representatives of deciduous temperate climate species, which drop their leaves at the end of the reproductive cycle and, consequently, entering in dormancy.

It is fruit maturity when the basic cycle of growth of a plant is completed. In annual crops the fruit would be the beginning of plant senescence and its subsequent death, whereas in perennial crops the duration of life consists of repeated growth cycles (Okubo 2000) marked in temperate plants for each cycle of dormancy. For a better understanding, therefore, a growth cycle in a quince, for example, corresponds to a life cycle in an annual plant.

2. Biological Objectives of Dormancy in Deciduous Plants

For the occurrence of a new vegetative cycle, it is necessary to expose the plant to a period of low temperatures so that dormancy is overcome and bud sprouting is in fact effective.

The usual sequence of events that lead these plants to winter dormancy is marked by pre-dormancy, initiated by the gradual reduction of the daily photoperiod as well as the reduction of average temperatures (Brighenti 2013).

The low autumnal and winter temperatures that can occur in subtropics and are more pronounced in the typically temperate regions are crucial for plant induction to dormancy, being one of the fundamental factors for triggering this physiological process.

The dormancy has its control by factors of hormonal nature from a regulation where the phytohormones and the growth inhibitors are balanced. The edaphoclimatic conditions, mostly the temperature, affect the plants marking the entrance and the exit of the dormancy period.

3. Dormancy versus Other Physiological Events

Before studying the induction and overcoming of dormancy itself, a great distinction of physiological processes becomes indispensable. There are processes that induce the plant to a state of dormancy and processes with important functions to the tolerance of the plants to the freezing and even the dehydration. Temperate zone plants are exposed annually to freezing temperatures each winter. Their ability to survive depends on an evolved mechanism by which plants enter into a state of dormancy and also develop resistance to cold, that is an acclimatization to cold beginning in autumn.

During the winter months, while the buds are totally dormant, plant tissues are characterized by a certain physical resistance. After the return of warmer temperatures in the spring, these changes culminate in overcoming dormancy. Thus, the process of beginning to overcome plant dormancy in the annual growth cycle is superimposed on a seasonal development and the loss of resistance to cold, triggering physiological and molecular changes associated with the regulation of dormancy (Arora 2003).

4. Induction to Dormancy

The dormancy is caused by the regulation imposed on the progressive processes of growth in the various stages, such as vegetative growth, flower formation and anthesis, and may present morphological changes. The induction of dormancy can therefore be defined as a change of the meristem that cease their growth for a time or that initiate special organs rather than produce shoots.

The modification of specific physiological processes that determine the plant growth paralysis and the acquisition of cold tolerance is a reflection of the synchronization of plant development with the existing environmental conditions, thus inducing the plant to dormancy. The environmental factors, plant-related factors and their interaction are determinant in the manifestation and control of dormancy in temperate fruit trees (Hawerroth 2010).

The entry into dormancy is also related, chronologically, to the remobilization of the nutrients of the senescent leaves to the organs that act as reservoirs of deciduous plants, such as trunks, roots and branches.

5. Phases or Types of Dormancy

Types or classes of dormancy may interact with each other chronologically, and when this happens, they constitute phases within the total dormancy period, involving some common control mechanisms.

Ecodormancy usually occurs after endodormancy in late winter and spring. However, in some cases, it may also occur before endodormancy. It is imposed by unfavourable temperatures to plant growth, water stress and nutritional deficiency, forcing the plant to keep its structures dormant until the conditions become again favourable.

The period of ecodormancy tends to be prolonged until the buds are exposed to temperatures sufficiently high for their opening, which in western Europe usually occurs between April and May (Jackson 2003) and in southern Brazil between September and October.

Ecodormancy is also observed in summer in subtropical climate species with persistent leaves, due to temporary water deficiency and in cold areas due to low temperatures. These plants will also initiate a new flow of growth once the limiting factor is extinguished (Salaya 2009).

Paradormancy is equivalent to correlative inhibition or apical dominance. It predates endodormancy, being caused by factors extrinsic to the affected bud (structure), influenced by biochemical signals coming from other organs of the plant, causing the inactivation of the floral or vegetative meristem such as that imposed by the leaves in the axillary buds. This phenomenon is also observed in lateral buds of branches due to the dominance of the terminal bud. If the inhibitory organ is eliminated, the affected structure will grow and develop again. Paradormancy is usually more strong in young plants.

Endodormancy, also called deep dormancy or winter dormancy, which is the genuine dormancy that characterizes woody plants in temperate zones, has been the subject of many studies that have shown the enormous complexity of this phenomenon (Campoy 2011). At this stage, dormancy is induced by internal factors of the affected tissue/structure as the plant responds to the decrease in temperature and / or photoperiod.

Therefore, ecodormancy and paradormancy are characterized by a "ceasing" of growth forced by physiological and/or environmental factors, whereas endodormancy is induced by plant intrinsic factors.

The term "entrance to dormancy" has been described by some authors like Westwood (1982) as the transition from ecodormancy to deep dormancy or endodormancy.

There are very large variations in the time of entry into endodormancy and also in the depth patterns of this dormancy between different cultivars, types of buds and buds of vertically inclined branches. This suggests that endodormancy is not triggered and developed in response to a single dominant environmental signal, which would operate directly on the buds (Jackson 2003), but because of multiple factors inducing biochemical and physiological responses.

There are two possible environmental causes that induce paradormancy (equivalent to the correlative inhibition caused by apical dominance or inhibition) for endodormancy (or dormancy) in the plants and that simultaneously initiate acclimatization to the cold. They are: the gradual decrease of the photoperiod and the decrease of the temperature (Arora 2003).

After the endodormancy, the appearance of new branches and leaves may be inhibited by adverse climatic conditions, mainly by cold, initiating ecodormancy (Milech 2015). However, as a result of an increase in photoperiod and temperature rise, the development proceeds rapidly and the plant will return to its development after the dormancy period.

In Figure 1, the stages of dormancy are exposed in a cycle within the four seasons, whose particular characteristics are usually well expressed in typically temperate regions, and moderately expressed in the subtropics. As mentioned above, dormancy classes may or may not succeed each other, thus composing phases of the total dormancy period.

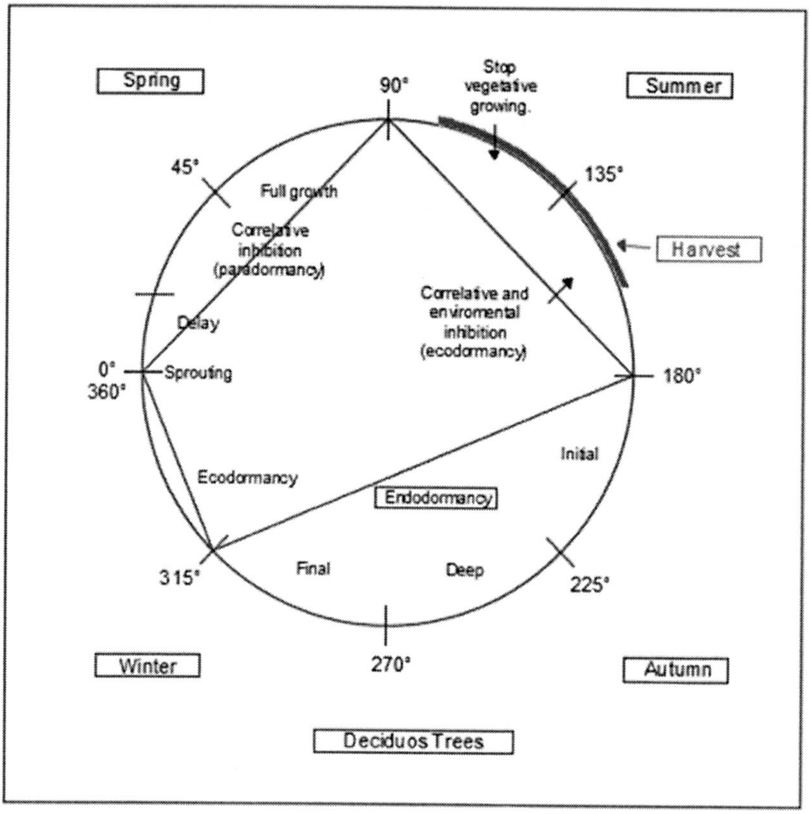

Figure 1. Annual growth cycle of deciduous temperate fruits. (Adapted from Salaya (2009) by the authors).

In 360°, in a perfect circle, some processes of dormancy breaking are represented: 0° represents sprouting, and intense vegetative growth until the 90°; passing through a possible paradormancy to the 45°, the beginning of the harvest of fruits, around the 135°; the stop of vegetative growth of the summer period in some previous degrees. In this same time represented in degrees, it is still observed the correlative inhibition (paradormancy) and in some cases the ecodormancy, being associated or not. Then, endodormancy begins with the arrival of autumn (180°), presenting a deeper and sharper form in a 45° interval (from 225° to 270°), finished at 315°, marked by a possible ecodormancy until the environmental and/or nutritional conditions are again favourable to sprouting, occurring the break of dormancy again, ending the cycle at 360°.

6. Overcoming Dormancy

The low temperatures act in a double process: initially contributing to the stoppage of growth, cold acclimatization and induction to dormancy, later they act to overcome this state. The amount of cold from induction to the overcoming of deep dormancy is called chilling requirement, which is variable between different species and cultivars. After complete compliance with the chilling requirement of a given species/cultivar to overcome endodormancy, there is a need for temperatures higher than those effective for cold accumulation, in order to accelerate the metabolic activities in the meristematic tissues of the buds and thus trigger sprouting (Hawerroth 2010). This is a consequence of the process of overcoming dormancy.

Until a few years ago, the chilling requirement of plants was measured relative to temperatures below 7.2°C (45°F). However, nowadays, the chilling hours below this temperature are considered only as a reference value, because the different temperatures can present varying effective values of accumulated amount of chilling that will take the plant out of dormancy.

It is also important to consider that under natural conditions temperatures are presented as a cycle, and in this cyclic form, temperatures below 0°C (32°F) are not effective, so that the physiological processes of accumulation of chilling hours is stopped, while temperatures above 7.2°C actually produce some chilling accumulation (Brighenti 2013).

In some cases, where chilling requirements are not met due to winters with temperature irregularities or even regions with relatively mild winter weather, a number of anomalies may occur. These anomalies, in relation to sprouting and flowering, have been described for several years. But even after sprouting and flowering, other anomalies may persist or manifest during the crop cycle, such as developmental paralysis of the branch, small leaves, small and flattened fruit in pomes, reduced density of spurs for the next year and low effective fruit set in temperate plants (Petri & Leite 2004).

In addition, it is common, in some years, the non occurrence of the effective overcoming of dormancy in plants, a phenomenon caused by insufficient chilling accumulation. In such cases, it is necessary to use chemical compounds for the artificial overcoming of dormancy, to obtain a sprouting with abundance and uniformity.

Many chemical compounds such as mineral oil, dinitro-ortho-cresol, thiourea, potassium nitrate, calcium nitrate, gibberellic acid and kinetin (Botelho 2003; Brighenti 2013), thidiazuron, hydrogenated cyanamide, sodium pentachlorophenolate (Brighenti 2013), all in their various trade names, are cited as effective ingredients in the overcome of dormancy of many species of fruit trees. Such compounds may partially replace the need for chilling and stimulate an early and more uniform bud sprouting (Botelho 2003).

These chemicals, as well as plant regulators, are a very useful tool that the fruit grower has at his disposal. However, their correct use results from the in-depth knowledge of plant physiology and the mechanism of action of each product. In order to extract the best results from its application, it is always recommended to carefully observe the ideal season and the indicated concentrations (Brighenti 2013), since its efficacy or a possible toxicity depend only on the dose and the state of dormancy depth (Sozzi 2007).

Final Considerations

The multiplicity of facts and phenomena that occur in the plant in the pre-dormancy period until its overcoming, subject to hormonal balances, climatic and soil conditions, and added to anthropic intervention activities, such as cultural practices, are responsible for a new life cycle - growth and reproduction - of deciduous plants of temperate climate. At the end of the dormancy period there is an increase in the concentration of nutrients responsible for promoting the activity of genes and enzymes. Biosynthesis, as well as reserve mobilization and basal metabolism, are then resumed and cell division restarts gradually again.

References

Arora, Rajeev; Rowland, Lisa, J.; Tanino, Karen, k. Induction and Release of Bud Dormancy in Woody Perennials: A Science Comes of Age. American Society for Horticultural Science. *Hortscience,* Vol. 38, Aug. 2003. United States of America, 2003.

Botelho, Renato Vasconcelos; Pires, Erasmo José Paioli. Quebra induzida [Induced breakage]. *Revista Cultivar Hortaliças e Frutas*. Edição 21, agosto/setembro 2003.

Brighenti, Lívia Mattos. Dormência da pereira [Pear tree dormancy]. In: Rufato, L.; Kretzschmar, A. A.; Bogo, A. *A cultura da pereira*. 1. ed. Florianópolis: DIOESC., 247 p. 123 - 149. 2013. (Série Fruticultura).

Campoy, José Antonio; Ruiz, David; Egea, J. Dormancy in temperate fruit trees in a global warming context: a review. *Scientia Horticulturæ,* Vol. 130, Issue 2, Sep. 2011, Pages 357 - 372. DOI: 10.1016/ j.scienta.2011.07.011.

Hawerroth, Fernando José; Herter, Flavio Gilberto; Petri, José Luiz, Leite, Gabriel Berenhauser; Pereira; José Francisco Martins. Dormência em frutíferas de clima temperado [Dormancy in temperate fruit trees]. *Embrapa Clima Temperado* – Pelotas: 2010. 56 p.

Jackson, John E. Biology of aples and pears. Chapter 6. *The shoot system*. Cambridge University Press. 485 p. Cambridge, United Kingdom, 2003.

Milech, Chaiane Goveia. *Estimativas da necessidade em frio de genótipos de pessegueiros por modelos matemáticos* [*Estimates of the cold need of peach genotypes by mathematical models*]. Dissertação de Mestrado em Agronomia. Universidade Federal de Pelotas; 97 f. - Pelotas, 2015. Documentos da Biblioteca Digital da Embrapa Clima Temperado – In: http:// ainfo. cnptia. embrapa. br/ digital/ bitstream/ item/ 137480/ 1/ DISSERTACAO-MESTRADO-CHAIANE-MILECH.pdf. Found in: July 25th, 2018.

Okubo, Hiroshi. Dormancy in plants: from whole plant behaviour to cellular control: Chapter 1. *Growth Cycle and Dormancy in Plants*. Cambridge: Cab Publishing, 2000.

Petri, José Luiz; Leite, Gabriel Berenhauser; Consequences Of Insufficient Winter Chilling On Apple Tree Bud-Break. *Acta Horticulturæ*, 662, 53 - 60 DOI: 10.17660/ActaHortic.2004.662.4. 2004.

Salaya, Gonzalo, F. Gil; El crescimento vegetativo. Latencia y letargo de yemas [The vegetative growth. Latency and bud lethargy]. In: *Fruticultura - El Potencial Productivo.* Ediciones Universidad católica de Chile. Santiago, Chile. Cuarta edición. Chile, 2009.

Sozzi, Gabriel, O.; Gariglio, Norberto, F.; Orell, María, Inés Figueroa De; Dormición em árboles frutales de hojas caducas [Dormancy in deciduous fruit trees]. Pg. 85 – 101. In: Sozzi, Gabriel, O. *Árboles frutales: ecofisiologia, cultivo y aprovechamiento*. Universidad de Buenos Aires. Argentina, 2007.

Viémont, Jean-Daniel; Crabbé, Jacques (Ed.). Preface. In: *Dormancy in plants: from whole plant behaviour to cellular control*. Cambridge: Cab Publishing, 2000.

Westwood, Melvin, N.; Latencia y adaptación al frio [Latency and adaptation to cold]. In: *Fruticultura De Zonas Templadas*. Oregon State University. Versión Española. Pg. 329 - 336. Ediciones Mundi-Prensa. Madrid, 1982.

In: Plant Dormancy
Editor: Renato V. Botelho

ISBN: 978-1-53615-380-4

Chapter 2

METHODS FOR ESTIMATING CHILLING REQUIREMENTS IN FRUIT CROPS

Thiago Marchi[1,*], Alberto Ramos Luz[2] and Renato Vasconcelos Botelho[3]

[1]Agricultural Research and Extension Company of the State of Santa Catarina (Epagri), Lajeado Grande, Santa Catarina State, Brazil
[2]Work Cooperative of Agronomy Professionals Ltda (Unicampo), Ipê, Rio Grande do Sul State, Brazil
[3]State University of the MidWestern of Paraná (Unicentro), Guarapuava, Paraná State, Brazil

ABSTRACT

The use of methods to estimate chilling requirement of temperate fruit trees are important tools for the prediction of physiological phase changes in the plant, as well as the decision making on the cultural management to be carried out in the orchards. Several methods can be used to estimate chilling requirement of temperate fruit trees. They can be basically divided into biological methods and temperature modeling methods. Biological

[*] Corresponding Author's Email: marchithiago@yahoo.com.br.

methods are those characterized by direct phenological observation (sprouting and flowering) of plant material (parts of plants or whole plants) submitted to controlled and distinct conditions during dormancy. In general, this method is used in experiments to characterize genotypes with respect to the chilling requirement or for studies of temperature modeling, evaluating the behavior of the buds in different situations. On the other hand, the methods of temperature modeling have the objective of measuring the amount of chilling accumulation during the winter period and, from the characterization of the need of the different cultivars, it is possible to predict the time of budbreak. Although widely used and an important tool in the management of fruit trees, this second method presents a very variable performance in certain conditions and crops, since the methodologies developed, for the most part, were based on empirical observations and not on physiological processes. Advances in studies on the mechanisms involved in tree dormancy control contribute to the development of more accurate models, increasing efficiency in predicting the physiological development of trees and improving information for decision-making in orchards management.

Keywords: dormancy, budbreak, flowering, temperate climate

1. Introduction

Dormancy is a complex biological phenomenon, in which great changes are verified in the plant metabolism, in order to acquire resistance to environmental conditions unfavorable to the development of plants (Hawerroth et al. 2010; Campoy et al. 2011). At this stage, plants adopt morphological and physiological mechanisms that allow survival at low temperatures in the winter.

Several factors related to the plant, the environment and the interaction of these factors may influence the dormancy process. It is known that photoperiod, temperature, rainfall and light affect this process, however, the temperature is considered the main climatic element that influences the plant dormancy (Hawerroth et al. 2010; Erez and Couvillon 1987; Faust et al. 1997; Erez 2000). The low temperatures act in two processes: initially they contribute to the stoppage of growth, acclimatization to cold and induction to dormancy and, later, they act in the overcoming of dormancy (Hawerroth et al. 2010).

During the fall begins a process in the metabolism of temperate climate trees that we call endodormancy. At this stage the low temperatures induce physiological changes that inhibit the development of meristems, resuming their development after exposure to a sufficient amount of low temperatures, called chilling requirement (Barros et al. 2012). Once this requirement is satisfied, ecodormancy occurs, where the development of flowering buds is dependent on the accumulation of high temperatures for a certain period, which we call the heat requirement. This second stage is finished with the anthesis of 50% of the flowers (Maulión et al. 2014). The chilling and heat requirement are specific characteristics of each genotype and play a fundamental role in the selection of suitable materials for each geographic region, since they are parameters that must be met in order to overcome dormancy and stability of flowering (Bassi et al. 2006).

Thus, in order for an orchard to achieve adequate yields, the genotypes should be chosen to coincide with the amount of chilling accumulation from the place where it will be implanted, so that an adequate sprouting occurs after the winter period and the flowering is delayed enough to avoid frost damage (Okie and Blackburn 2011). Failure in choosing appropriate cultivars to the implant site could have catastrophic consequences for the growers, since orchards have, in general, a long productive life and a high cost of establishment (Luedeling et al. 2009).

Therefore, the understanding of phenology provides not only information about the performance and adaptation of genotypes, but also reliable information for selection in breeding programs and for the guidance of growers in the use of crop management techniques that improve sprouting and/or prevent cold damage (Maulión et al. 2014). In this sense, some methodologies can be used to predict the sprouting time of different genotypes, and then define strategies for the orchard establishment and management of temperate fruit crops.

2. Methods for Estimating Chilling Requirements

Due to the need to elucidate the physiological mechanisms that govern dormancy in temperate fruit trees, as well as to evaluate the influence of environmental factors on this process, especially for determination of chilling requirement of different cultivars, several methods were developed in the study of dormancy (Hawerroth et al. 2010). These methods can be divided into biological methods and temperature modeling methods (Maulión et al. 2014).

2.1. Biological Methods

The biological methods are characterized by the use of plant material in evaluations, aiming to elucidate the various mechanisms related to dormancy (Hawerroth et al. 2010). In general, these methods are used in experiments that require the use of a large number of observations, as in breeding programs, searching for the selection of genotypes with a certain chilling requirement or as in modeling of dormancy, aiming to verify the effects of different temperatures on dormancy process (Anzanello et al. 2014). The use of plant material is essential in studies with higher accuracy, thus requiring that the plant material under evaluation represent trustingly the field situation (Hawerroth et al. 2010).

Biological methods are characterized by direct phenological observation (sprouting and flowering) in plant material submitted to controlled and distinct conditions during dormancy (Dole 2001). In the field, there is no control over solar radiation, precipitation and temperature, and it is difficult to determine with precision the chilling requirement of fruit species, due to the complexity of interactions that environmental and plant factors exert on tree dormancy (Dennis Junior 2003).

There are many variations of biological methods, and whole plants or only parts of them can be used, depending on the specific research objectives. Among the biological methods, we can mention the testing of isolated node cuttings, use of detached shoots, use of potted trees and use of grafted branches (Hawerroth et al. 2010).

The isolated node cuttings test is a methodology that was developed at the University of Clermont-Ferrand in France and is most frequently used to characterize the level of dormancy over the winter period (Anzanello et al. 2014). Although there is a biochemical method (nucleotide test) to characterize the level of dormancy, the biological test of isolated node cuttings is the only one that quantifies the depth of dormancy, is practical and has a good correlation with the biochemical tests (Crabbé and Barnola 1996). However, in this case, the plant material presents low durability, even when planted in phenolic foam, due to the lower supply of reserves, greater exposure of the internal tissues and greater hormonal imbalance, as a function of the ethylene and abscisic acid induced by the cuts in tissues (Dennis Junior 2003).

The detached shoot method is preferable for modeling and sprouting prediction studies, by partially maintaining the interrelations between plant tissues and, consequently, bringing the responses closer to the real conditions (Anzanello et al. 2014). According to Dole (2001), the detached shoot method is based on the principle of the correlative inhibition that one bud has over another, evaluating the paradormancy and endodormancy simultaneously. However, when isolated node cuttings, containing a single bud are used, the paradormancy is eliminated, only endodormancy is acting (Champagnat 1983). The sprouting difference observed between the two methodologies indicates the intensity of the paradormancy. Many methodologies for evaluating dormancy recommend testing isolated nodes (Herter 1992; Balandier 1992), however, if the interaction between the buds is expressive, there may be problems in the interpretation of the results and in the evaluation of the physiological conditions of the buds in the trees (Anzanello et al. 2014).

Anzanello et al. (2014) describes the biological method of detached shoots, packed with plastic wrap in upward vertical position during cold treatments, and conditioned in phenolic foam during the induction of sprouting in warm conditions, is a suitable method to evaluate the state of dormancy of buds in apple trees for modeling studies. The use of detached shoots held vertically corresponds to the condition that most closely approximates the natural state of the buds in apple branches, which is important for modeling the dormancy process.

Other methods can be used to evaluate bud dormancy, such as the use of whole plants or grafted branches, which allow to evaluate interactions between buds and other tissues or organs (Hawerroth et al. 2010). However, for Camelatto et al. (2000) and Hawerroth et al. (2010) the use of potted plants for studies of dormancy is limited by the demand for large space for its realization, raising costs due to the need for large climatized growth chambers. In addition, during the juvenile period, which tends to vary from 3 to 5 years (Simão 1971), there is no floral buds formation. In peach trees, the method of grafted branches is considered promising to avoid juvenile problems in chilling requirement studies (Wagner Junior et al. 2006; Chavarria; Herter and Raseira 2009). The method consists of grafting productive branches on rootstocks previously established in soil or substrate, and the treatments are applied after the complete welding at the grafting point. Its advantages are the ability to maintain vegetative material for longer periods than by using detached shoots due to less dehydration and less need for space in relation to potted plants. However, the grafting method still requires a much more space in relation to the detached shoots method, especially when it is necessary to use several thermal regimes.

2.2. Methods of Temperature Modeling

As low temperatures are the main climatic element that influences the dormancy process, several works were done to try to model this factor (Luedeling 2012), as we can see in the Table 1. These models consider different ways of measuring the chilling amount accumulated during the period when the plant is dormant. After reaches the necessary amount of chilling, the buds are able to overcome this stage and, under suitable environmental conditions, they would begin the sprouting stage.

Table 1. Overview of major chilling models and comparison with regard to the inclusion of major scientific insights into temperature effects on chill accumulation

Model names and authors	Time step	Differentiall temperature weights	Continuous weights	Chill negation by heat	Limit to chill negation	Enhancement by moderate temps.	Two-phase chilling
Chilling Hours Model (Weinberger 1950)	H	-	-	-	-	-	-
Utah Model (Chill Units) (Richardson et al. 1974)	H	+	-	-	-	-	-
North Carolina Model (Shaltout and Unrath 1983)	H	+	-	+	-	-	-
Anderson and Seeley 1992	H	+	+	+	-	-	-
Positive Utah Model (Linsley-Noakes and Allan 1994)	H	+	-	+	+	-	-
Modified Utah Model (Linvill 1990)	H	+	+	+	-	-	-
Regional Models in Georgia and Flórida	M	-	-	-	-	-	-
Chmielewski et al. (2011)	D	-	-	-	-	-	-
Legave et al. (2008)	D	+-	+-	-	-	-	-
Cesaraccio et al. (2004)	D	+	+	-	-	-	-
Dinamic Model (Chill Portions) (Fischman et al. 1987a, 1987b)	H	+	+	+	+	+	+

+ indicates that the respective characteristic is included in the model; – indicates that it is not included; ± means that different versions exist, with only some including the characteristic.

Model characteristics are temperature step (h = hourly, d = daily, m = monthly), differential weighting of different temperature ranges, continuous (as opposed to step-wise) distribution of weights, negation of earlier chill by high temperatures, a limit to how much chill can be negated, enhancement of earlier chill by moderate temperatures and an assumed two-step process of chill accumulation.

The authors given in the table are not necessarily the model developers.

The first model to be identified is the Chilling Hours model, also known as the Weinberger Model (Weinberger 1950), which expresses the amount of time (in hours) occurring in the winter period, where temperatures are in the range of 0 to 7.2°C; considering this temperature range as effective for overcoming the dormancy process. The Weinberger method (1950) is still widely used, due to the ease of calculation, mainly in the characterization of cultivars in breeding programs, although it is a low precision method when applied in regions with low chilling accumulation (Linsley-Noakes, Allan and Matthee 1994). According to Erez (2000), the principles that the chilling requirement is quantitative and can be measured by exposure to a certain number of chilling hours and that there is a critical temperature which below it is effective in dormancy are the main contributions established by the Weinberger method (1950).

Subsequently, a significant advance in the modeling of environmental conditions was established with the discovery that high temperatures have a negative effect on the winter chilling accumulation of dormant trees (Overcash and Campbell 1955). One example is the Utah Model (Richardson et al. 1974), which considers different weight scales for different temperatures, including negative weights for temperatures above 15.9°C. This model was developed for peach trees in Utah, USA.

After the Utah model, several methodologies adapted to different regions and fruit species were created (Gilreath and Buchanan 1981; Shaltout and Unrath 1983; Linvill 1990; Anderson and Seeley 1992; Linsley-Noakes and Allan 1994; Warmund and Krumme 2005). These models consist basically of the conversion of hourly temperatures into chilling units (CU), and the units are accumulated daily until reaching a total that theoretically corresponds to the end of the dormancy phase of a given plant tree cultivar (Hawerroth et al. 2010).

The third approach in methods of temperature modeling for purposes of characterizing tree dormancy is known as Dynamic Model (Fishman et al. 1987a, 1987b). This model is based on the assumption that the chilling accumulation is the result of a two-step process. The first part of the process consists in the formation and reversible destruction of the intermediate thermal product, where low temperatures promote the reaction, while high temperatures favor the reaction reversion. When a critical level of the intermediate product is reached, it is irreversibly transferred to the second portion of the model, where the accumulation of factors for sprouting induction occurs, not being degraded or reduced by the effect of subsequent high temperatures (Allan 2004). This model is the only one among the common models that can explain the observed negative effect of high temperatures (Vegis 1961; Thompson et al. 1975; Couvillon and Erez 1985), the apparent limit to how much chill can be reversed (Erez et al. 1979), and the chill-enhancing effect of moderate temperatures when cycled with cool conditions (Erez and Couvillon 1987).

For Luedelling (2012) an important difference between the Dynamic Model and the previous approaches is the importance given to the sequence of temperatures during the cold season. According to the Chilling Hours and Utah models, similar temperatures always have exactly the same effect regardless of when they occur. In the Dynamic Model, several processes interact and the production of a chilling quantity depends on the existence of a certain amount of the intermediate product. Similar temperatures at different times of the season may therefore have very different effects on the chilling accumulation. Zhang and Taylor (2011) indicate that the Dynamic Model has the quality that the measurement of the chilling quantity at different moments of the winter period are not homogeneous, as opposed to the homogeneous nature of the other models. Several authors argue that the structure of the Dynamic Model, or its homogeneous rate of cold accumulation, makes it the most plausible among the common models (Erez and Fishman 1990; Allan et al. 1995, 1997; Perez et al. 2008; Luedeling et al. 2009a, 2009b, 2009c, 2011a, 2011b; Darbyshire et al. 2011; Luedeling and Brown 2011).

3. Problems Related to the Use of Methods for Temperature Modeling

Although all common horticultural models have been successful to a certain extent in guiding orchard management and selection of cultivars, they are quite different and not all equally reliable (Luedeling 2012). Thus, there are no universal models capable of accurately predicting the time of dormancy overcoming for several species and for various environmental conditions (Cesaraccio et al. 2004).

The variable efficiency of chilling modeling methods may be related to several factors. According to Erez (2000), the Chilling Units model exhibit satisfactory performance under typically temperate conditions, but these methods are flawed when used to predict the end of dormancy under subtropical conditions. According to Luedeling (2012) most models lack precision, and for regions of milder winter temperatures, the plant behavior of overcoming dormancy has not yet been explained at all. However, studies of trees grown at marginal sites are very important, since they can produce important information about the empirical relationship between temperature and chilling accumulation. For Finetto (2004), most of the elaborated models consider only the temperature factor, although there is evidence that the light and precipitation are also factors that affects overcoming dormancy, being one of the reasons for the divergence in the results of the chilling requirement of different plant cultivars. Fuchigami and Wisniewiski (1997) also consider that the buds dormancy can be altered by cultural practices and the environment conditions, so the misunderstanding of the way in which these factors influence the dormancy applied in the models can reduce the accuracy of these studies.

Moreover, most models are based exclusively on observations in the field, while systematic controlled environment experiments seem to have been used only for the Dynamic Model (Luedeling 2012). This is also the only one of the commonly used models that can be called 'process-based' (as opposed to purely empirical) models, although the development of models has been based on hypothetical processes (Fishman et al. 1987a, 1987b), rather than processes with a solid scientific basis.

In addition to the poor availability of information regarding the chilling requirement of the species and/or cultivars, the existing information is often contrasting, since the chilling requirement is often expressed in different measures, making it difficult to use (Hawerroth et al. 2010). For this reason, the introduction of new cultivars, in regions with different climatic conditions from where they originated, may lead to problems caused by non-fulfillment of the chilling requirement or earlier flowering (Valentini et al. 2004).

Luedeling (2012) evaluated several studies comparing temperature modeling methods with tree phenology. According to this author, some models presented good performance, being able to predict dates of flowering within a few days of the actual flowering dates. However, as the variation in observed flowering dates is not very large, all model combinations have left a large part of this variation unexplained. This situation, even when the models were calibrated with large datasets observed, raises some doubts about the mathematical structure of the models and the assumptions underlying the different modeling approaches, considering that most of the research available is based on empirical observations, not on the complete understanding of plant physiology. Some models still performed reasonably well in reproducing observed flowering patterns, but their adequacy to climate change projections should be questioned.

Conclusion

The overcoming of dormancy is a process that occurs due to the interference of several factors related to the tree and the environment. Due to a large gap in the knowledge of tree dormancy mechanisms and the fact that the temperature modeling methods available so far have a purely empirical basis, we should not be surprised if they prove to be inaccurate at times.

It is believed that increasing advances in studies on molecular and physiological control mechanisms of tree dormancy may contribute to the development of more accurate models (Campoy et al. 2011), increasing efficiency in predicting the physiological development of plants and improving information for decision-making in the cultural management to be carried out in the orchards.

The Dynamic Model currently appears to be the best model in terms of accuracy, however, the Chilling Hours approach to chilling quantification is the most widely used in several parts of the world, mainly for farmers, due to its lower complexity.

Studies on the behavior of temperate fruit plants at sites with lower winter chilling accumulation, such as in subtropical regions, are of great importance for the development of more accurate modeling of environmental conditions (considering large temperature oscillations in the winter period), for the development of cultural practices to alleviate the lack of chilling and for the development of technologies for areas that will suffer from climate change.

REFERENCES

Allan, P. 2004. "Winter chilling in areas with mild winters: its measurement and supplementation." *Acta Horticulturae* 47:52-662.

Allan, P., Linsley-Noakes, G. C., Holcroft, D. M., Brunette, S. A., Burnett, M. J., CathcartKay, A. 1997. Kiwifruit research in a subtropical area. *Acta Horticulturae* 37:42-444.

Allan, P., Rufus, G., Linsley-Noakes, G. C., Matthee, G. W. 1995. Winter chill models in a mild subtropical area and effects of constant 6°C chilling on peach budbreak. *Acta Horticultae* 9:17-409.

Anderson, J. L., Seeley, S. D. 1992. Modelling strategies in pomology: development of the Utah Models. *Acta Horticulturae* 297:306-313.

Anzanello, Rafael; Filho, Flávio Bello; Santos, Henrique Pessoa Dos; Bergamaschi, Homero; Marodin, Gilmar Arduino Bettio 2014. Biological methods to evalute the sprouting of buds in apple trees to dormancy modeling. *Semina Ciências Agrárias* 1163:1176-35.

Balandier, P. 1992. *Study of the dinamic of the growth and development of buds of some peach cultivars at diverse altitudes in tropical climate of l'île de la.* Thesis (PhD in Plant Physiology Physiologie), University Blaise Pascal.

Barros, P. M.; Gonçalves, N.; Saibo, N. J. M.; Oliveira, M. M. 2012. Cold acclimation and floral development in almond brud break: insights into the regulatory pathways. *Journal of Experimental Botany* 5485:4596-63.

Bassi, D., Bartolini, S., Viti, R., 2006. Recent advances on environment and physiological challenges in apricot growing. *Acta Horticulturae* 23:31-717.

Camelatto, D.; Nachtigall, G. R.; Arruda, J. J. P.; Herter, F. G. 2000. Effects of temperature fluctuations, chilling hours and growth regulators in floral abortion in pear trees buds. *Revista Brasileira de Fruticultura* 111:117-22.

Campoy, J. A.; Ruiz, D.; Egea, J. 2011. Dormancy in temperate fruit trees in a global warming contexto: a review. *Scientia Horticulturae* 357:372-130.

Cesaraccio, C., Spano, D., Snyder, R. L., Duce, P., 2004. Chilling and forcing model to predict bud-burst of crop and forest species. *Agricultural and Forest Meteorology* 1:13-126.

Champagnat, P. 1983. Bud dormancy, correlation between organs, and morphogenesis in woody plants. *Fiziologiya Rastenii* 458:471-30.

Chavarria, G.; Herter, F. G.; Raseira, M. C. B. 2009. Effect of mild temperatures on bud breaking dormancy in low and medium chill peaches. *Ciência Rural* 2016:2021-39.

Chmielewski, F. M., Blümel, K., Henniges, Y., Blanke, M., Weber, R. W. S., Zoth, M., 2011. Phenological models for the beginning of apple blossom in Germany. *Meteorologische Zeitschrift* 487:496-20.

Couvillon, G. A; Erez, A. 1985. Effect of level and duration of high temperatures on rest in the peach. *Journal of the American Society for Horticultural Science* 579:581-110.

Crabbé, J.; Barnola, P. A 1996. New Conceptual Approach to Bud Dormancy in Woody Plants. In: *Plant Dormancy: Physiology, Biochemistry and Molecular Biology*, edited by Lang, G. A., 83-113. New York: CAB international.

Darbyshire, R., Webb, L., Goodwin, I., Barlow, S., 2011. Winter chilling trends for deciduous fruit trees in Australia. *Agricultural and Forest Meteorology* 1074:1085-151.

Dennis Junior, F. G. 2003. Problems in standardizing methods for evaluating the chilling requirements for the breaking of dormancy in buds of woody plants. *Hort Science* 347:350-38.

Dole, J. 2001. Standardizing methods for evaluating the chilling requirements for the breaking of dormancy in bulbs, corms, and tubers. *Hort Science* 341:346-38.

Erez, A. 2000. Bud dormancy: phenomenon, problems and solutions in the tropics and subtropics. In: *Temperate Fruit Crops in Warm Climates*. Erez, A., 17-48. The Netherlands: Kluwer Academic Publushers.

Erez, A., Couvillon, G. A., Hendershott, C. H., 1979. Effect of cycle length on chilling negation by high temperatures in dormant peach leaf buds. *Journal of the American Society for Horticultural Science* 573:576-104.

Erez, A.; Couvillon, G. A. 1987. Characterization of the influence of moderate temperatures on rest completion in peach. *Journal of the American Society for Horticultural Science* 677:680-112.

Erez, A; Fishman, S. 1990. The dynamic model for rest completion in peach buds. *Acta Horticulturae* 165:174-276.

Faust, M.; Erez, A.; Rowland, L. J.; Wang, S. Y.; Norman, H. A. 1997. Bud dormancy in perennial fruit trees: Physiological basis for dormancy induction, maintenance and release. *Hort Science* 623:629-32.

Finetto, G. A. 2004. Investigation of the effects of light and rain on overcoming of dormancy in some apple cultivars. *Acta Horticulturae* 93:100-662.

Fishman, S., Erez, A., Couvillon, G. A., 1987a. The temperature dependence of dormancy breaking in plants-computer simulation of processes studied under controlled temperatures. *Journal of Theoretical Biology* 309:321-126.

Fishman, S., Erez, A., Couvillon, G. A., 1987b. The temperature dependence of dormancy breaking in plants-mathematical analysis of a two-step model involving a cooperative transition. *Journal of Theoretical Biology* 473:483-124.

Fuchigami, L. H.; Wisniewiski, M. 1997. Quantifying bud dormancy: physiological approaches. *Hort Science* 618:623-32.

Gilreath, P. R., Buchanan, D. W., 1981. Rest prediction model for low-chilling Sungold nectarine. *Journal of the American Society for Horticultural Science* 426:429-106.

Hawerroth, F. J., Herter, F. G., Petri, J. L., Leite, G. B. and Pereira, J. F. M. 2010. *Dormancy in deciduous fruit trees*. Pelotas: Embrapa Clima Temperado.

Herter, F. G. 1992. *Dormance of buds and phenology of some apple trees cultivars: effect of the temperature and interaction with genotipe*. Thesis (PhD in Plant Physiology), University of Blaise Pascal.

Legave, J. M., Farrera, I., Almeras, T., Calleja, M., 2008. Selecting models of apple flowering time and understanding how global warming has had an impact on this trait. *Journal of Horticultural Science and Biotechnology* 76:84-83.

Linsley-Noakes, G. C.; Allan, P.; Matthee, G. 1994. Modification of rest completion models for improved accuracy in South African stone fruit orchards. *Journal of Southern African Society for Horticultural Sciences* 13:15-4.

Linsley-Noakes, G. C., Allan, P., 1994. Comparison of 2 models for the prediction of rest completion in peaches. *Scientia Horticulturae* 107:113-59.

Linvill, D. E., 1990. Calculating chilling hours and chill units from daily maximum and minimum temperature observations. *Hort Science* 14:16-25.

Luedeling, E., Blanke, M., Gebauer, J., 2009a. *Climate change effects on winter chill for fruit crops in Germany*. Erwerbs-Obstbau 81:94-51.

Luedeling, E., Brown, P. H., 2011. A global analysis of the comparability of winter chill models for fruit and nut trees. *International Journal of Biometeorology* 411:421-55.

Luedeling, E., Girvetz, E. H., Semenov, M. A., Brown, P. H., 2011a. Climate change affects winter chill for temperate fruit and nut trees. *PLoS One* 6(5), e20155. Accessed June 2, 2018. doi: https://doi.org/10.1371/journal.pone.0020155.

Luedeling, E., Kunz, A., Blanke, M., 2011b. More winter chill for fruit trees in warmer winters? *Erwerbs Obstbau* 145:155-53.

Luedeling, E., Zhang, M., Girvetz, E. H., 2009b. Climatic changes lead to declining winter chill for fruit and nut trees in California during 1950-2099. *PLoS One* 4, e6166. Accessed June 15, 2018. doi: https://doi.org/10.1371/journal.pone.0006166.

Luedeling, E., Zhang, M., Luedeling, V., Girvetz, E. H., 2009c. Sensitivity of winter chill models for fruit and nut trees to climate change. *Agriculture, Ecosystems & Environment* 23:31-133.

Luedeling, E. 2012. Climate change impacts on winter chill for temperate fruit and nut production: A review. *Scientia Horticulturae* 218:229-144.

Maulión, E.; Valentini, G. H.; Kovalevski, L.; Prunello, M.; Montia, L. L.; Daordenb, M. E.; Quaglinoc, M.; Cervigni, G. D. L. 2014. Comparison of methods for estimation of chilling and heat requirements of nectarine and peach genotypes for flowering. *Scientia Horticulturae* 112:117-177.

Okie, W. R., Blackburn, B. 2011. Increasing chilling reduces heat requirement for floral budbreak in peach. *Horticultural Science* 245:252-46.

Overcash, J. P., Campbell, J. A., 1955. The effects of intermittent warm and cold periods on breaking the rest period of peach leaf buds. *Proceedings of the American Society for Horticultural Science* 87:92-66.

Perez, F. J., Ormeno, N., Reynaert, J., Rubio B., B., 2008. Use of the Dynamic Model for the assessment of winter chilling in a temperate and a subtropical climatic zone of Chile. *Chilean Journal of Agricultural Research* 198:206-68.

Richardson, E. A.; Seeley, S. D.; Walker, D. R. 1974. A model for estimating the completion of rest for ‘Redhaven’ and ‘Elberta’ peach trees. *Hort Science* 331:332-9.

Shaltout, A. D.; Unrath, C. R. 1983. Rest completion prediction model for ‘Starkrimson Delicious’ apples. Journal of the American Society for *Horticultural Science* 957:961-108.

Simão, S. 1971. *Manual of fruit tree crops.* São Paulo: Editora Agronômica Ceres, 1971.

Thompson, W. K., Jones, D. L., Nichols, D. G., 1975. Effects of dormancy factors on growth of vegetative buds of young apple trees. *Australian Journal of Agricultural Research* 989:996-26.

Valentini, N.; Me, G.; Spanna, F.; Lovisetto, M. 2004. Chilling and heat requirement in apricot and peach varieties. *Acta Horticulturae* 199:203-636.

Vegis, A., 1961. Sprouting and vegetative development of the buds. Handbook of Plant Physiology. *Encyclopedia of Plant Physiology* 168:298-16.

Wagner Júnior, A.; Bruckner, C. H.; Pimentel, L. D.; Raseira, M. C. B. 2006. Evaluation of chilling requirement in peach through grafted twigs. *Acta Horticulturae* 243:246-713.

Warmund, M. R., Krumme, J., 2005. A chilling model to estimate rest completion of erect blackberries. *Hort Science* 1259:1262-40.

Weinberger, J. H. 1950. Chilling requirements of peach varieties. *Proceedings of the American Society for Horticultural Science* 122:128-56.

Zhang, J., Taylor, C., 2011. The Dynamic Model provides the best description of the chill process on 'Sirora' pistachio trees in Australia. *Hort Science* 420:425-46.

In: Plant Dormancy
Editor: Renato V. Botelho

ISBN: 978-1-53615-380-4

Chapter 3

ASSESSMENT OF DORMANCY INTENSITY IN BUDS OF TEMPERATE CLIMATE FRUIT SPECIES

***Ruy Inácio Neiva de Carvalho*[1,*] *and Luiz Antonio Biasi*[2]**

[1]Department of Agronomy, Pontifícia Universidade Católica, Curitiba, PR, Brazil

[2]Department of Crop Science and Plant Protection, Federal University of Paraná, Curitiba, PR, Brazil

ABSTRACT

Bud dormancy of temperate species is a natural and necessary cyclic process that occurs as part of a plant protection system against low temperatures during the winter. Several climatic factors are involved in this process, with emphasis on temperature, and internal factors such as phytohormones, carbohydrates, growth inhibitors, water content, enzymes and oxidative stress. Due to the interaction among these factors, the study of dormancy is complex and several methods have been developed to

* Corresponding Author's Email: ruy.carvalho@pucpr.br.

evaluate the presence and intensity of dormancy. In this chapter some of these methods will be described, discussing how they can be performed and their results. The Tabuenca's test is used for evaluation of floral buds dormancy. The single node cutting test is one of the most used to evaluate dormancy intensity. The dormancy index was created to more clearly characterize the dormancy intensity from the results of the single node cutting test. Biochemical tests can be performed to quantify various substances involved in the dormancy process, such as carbohydrates, enzymes and reactive oxygen species. The tetrazolium test is a newly developed test to rapidly quantify the respiratory activity of buds.

Keywords: dormancy, Tabuenca, bud burst, tetrazolium, ROS

1. Introduction

Dormancy is one of the most complex physiological processes occurring in temperate species. These species developed resistance mechanisms to survive in environments in which low temperatures during the winter could cause their death. Buds undergo through many biochemical and morphological changes over a long period of the year, the most striking modifications occurring in autumn and winter. The growth paralysis of the buds takes place by different effects and ways, which results in types of dormancy that occur sequentially, but also superimposed.

Endodormancy is the type of mechanism that prevents the buds of responding to favorable growth factors. The entrance, permanence and exit of the buds in this process has been studied for many years and a great knowledge has already been accumulated on biochemical, morphological and physiological aspects. Even so, several gaps still remain obscure and the interaction between substances and their broad functions in cellular metabolism further hamper the complete dormancy understanding.

The identification of a dormant bud or the intensity of dormancy in which this bud is found, is not possible visually. Therefore, methods have been developed to more safely confirm whether a bud is in endodormancy or ecodormancy and, the intensity of endodormancy.

A very common test is the testing of isolated buds, where it is possible to remove the paradormancy effect and estimate the dormancy intensity by the time the buds take to sprout and by the sprouting rate reached. From the parameters evaluated in this test, the dormancy index (DI), which in a weighted way, considering the greater or lesser contribution of each parameter, can characterize at what level of dormancy intensity the bud is found. Single node cutting test presents good results for vegetative buds, which tolerate a longer period of exposure to optimal growth temperatures. For the evaluation of flowering buds, the Tabuenca's test was developed, which allows the identification of endodormancy.

Biochemical tests have evolved and new discoveries are made regarding the metabolism involved in dormancy. Thus, studies on the content of the different types of carbohydrates and their dynamics, involving transport between the bud and stem tissues, its accumulation and degradation, has elucidated how the metabolism of carbohydrates occurs during the phases of endodormancy. Oxidative stress has been well studied and several enzymes have already been identified as responsible for regulating the balance of reactive oxygen species (ROS). This group of molecules has many metabolic functions, but also its involvement in the endodormancy releasing process, especially the effect of the hydrogen peroxide accumulation. Quantification of the enzymes and ROS present in the buds can characterize the dormancy process.

The respiratory activity of the buds undergoes great alterations with the entry into dormancy, being very reduced during the whole process, until the endodormancy is overcome. For the rapid quantification of the respiratory activity, the tetrazolium test for buds was developed, in which a few hours it is possible to complete the test and estimate the intensity of dormancy.

2. Tabuenca's Test

The Tabuenca's test, as the name suggests, was extensively used by the researcher Maria Concepción Tabuenca, from the Aula Dei Experimental Station in Zaragoza, Spain. Her studies were based on the method developed by Brown and Kotob (1957) from the University of California, who evaluated the growth of apricot, peach and pear flower buds during dormancy by determining fresh and dry weight of floral buds. The procedure adopted was to collect samples of 50 apricot and peach buds and 25 pear buds at intervals of 1 or 2 weeks, whose scales were removed and the buds dried to constant weight. As the fresh weight values presented the same dry weight tendency, the authors considered only the dry weight to discuss the results.

The first study by Tabuenca using this method was published in 1964 in the *Anales de la Estacion Experimental de Aula Dei*. This work was carried out to estimate the chilling requirements for dormancy release of 18 peach cultivars, 45 pear cultivars and 10 apricot cultivars. It was not possible to determine the number of chilling hours required by these species to overcome dormancy, but the test allowed the separation of the peach and pear cultivars in four groups according to the chilling requirements. With the apricot, the test allowed to estimate the chilling amount necessary for dormancy release of these cultivars (Tabuenca 1964). The periodical evaluations of dry weight of the buds allowed the elaboration of growth curves, characterized by the existence of three distinct periods (Brown and Kotob, 1957): an initial period of slow growth in autumn and early winter; a second period with gradual increase in dry matter accumulation and a third period with rapid growth in the weeks before flowering (Figure 1). The small weight gain in the transition period indicates that the resting phase is ending and when there is a rapid increase on flowering bud growth rates, it is indicative of the end of dormancy.

According to this method, the floral buds are collected and taken to the laboratory, where the scales and the peduncle are removed. After, the floral primordia are placed in an oven at 70°C and the dried weight is obtained. With this procedure, it is possible to evaluate the growth of the buds under natural field conditions. The branches of the plants containing the floral buds and take them to the laboratory, where they are kept at 20 ± 1°C for 10 days, removing, therefore, the effect of low temperatures during the winter, without conditions that induce the ecodormancy. After this period the floral buds are collected, cleaned and dried as previously described (Tabuenca 1964).

In continuity to her studies to estimate the chilling requirements for dormancy release of species of temperate climate, Tabuenca published another work in 1967 evaluating 8 cultivars of Japanese plum and 10 cultivars of European plum. In this work, the same methodology described previously was used, but with a change in relation to the time that the collected branches remain in the laboratory at 20 ± 1°C, which was of 7 days (Tabuenca 1967). This became the most used methodology in future works, which is detailed in Figure 2.

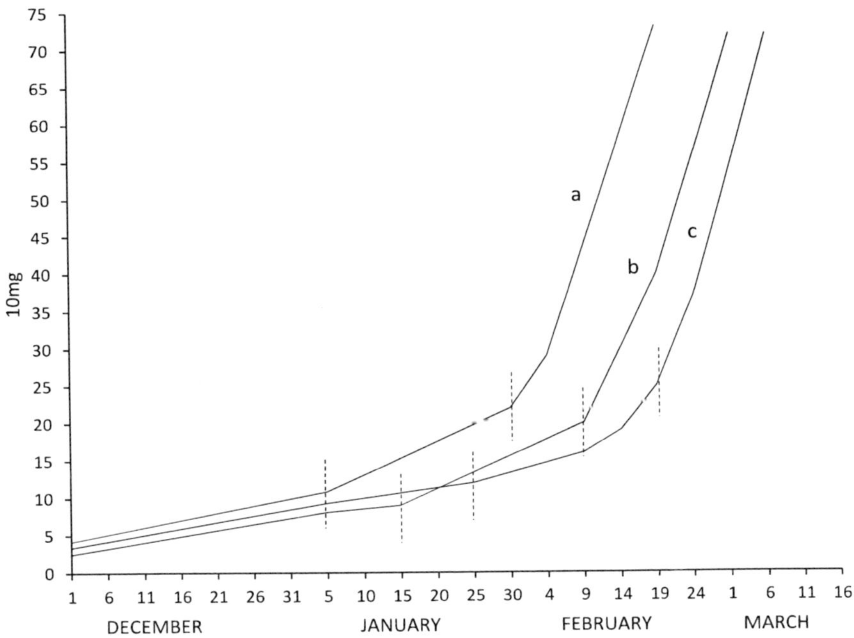

Figure 1. Evaluation of the floral bud dry weight from peach cultivars: a) Jerônimo; b) Duraznilla de Sástago and c) Mayflower. Adapted from Tabuenca (1964).

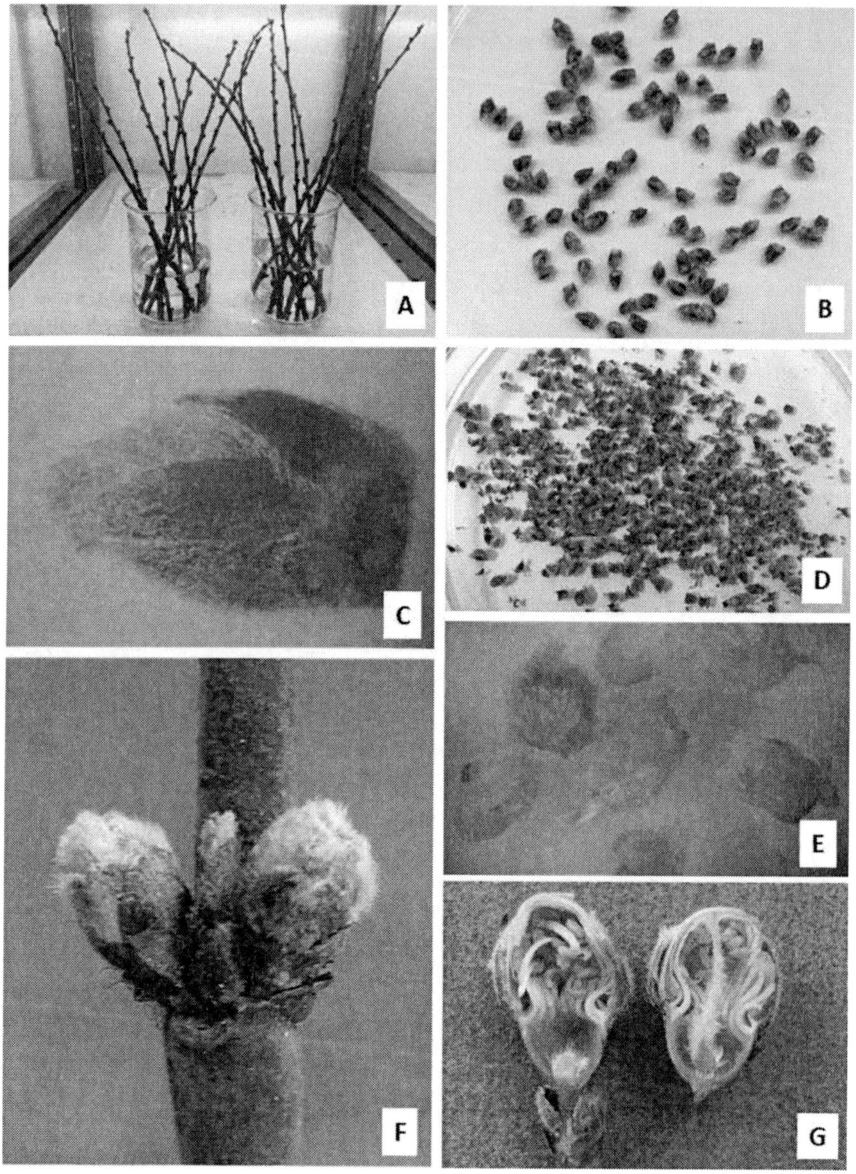

Figure 2. Tabuenca's test. A. Peach branches during the forcing test. B. Flower buds taken at the start of the test. C. Peach floral bud before forcing with scales. D. Scales removed from floral buds. E. Floral primordia without scales before forcing. F. Flower buds after the forcing period. G. Floral bud without endodormancy that presented growth after the period of forcing. Source: Biasi.

From these studies, Tabuenca published a long series of papers in the same scientific journal, estimating the chilling requirement and evaluating the phenological behavior of several cultivars in relation to the climatic conditions of Spain. In 1968, two studies were published, one with 58 pear cultivars, which were divided into four groups according to the cold requirement (Tabuenca 1968a) and the other with several apricot cultivars determining the cold requirement that varied within a range of 400 to 1200 hours of temperatures below 7°C (Tabuenca 1968b). Other similar investigations using floral buds were performed with peach (Tabuenca 1969, 1972b), almond (Tabuenca 1972a), nectarine (Tabuenca 1976a) and apricot (Tabuenca 1976b). However, in an experiment with 11 plum and 3 apple orchards of the Malling series, the vegetative buds were evaluated. In this case were collected in the field branches with about 40 to 50 cm in length that were kept at 20°C for 7 days with the base immersed in water. The percentage of swollen vegetative buds was evaluated weekly and the end of the dormancy period was considered when more than 30% of the vegetative buds were swollen after 3 weeks (Tabuenca 1976c).

Some variations in the method were introduced over time, such as determination of the fresh mass of whole buds before and after the "forcing test" (Guerriero et al. 2002). This procedure makes the method quicker and more practical, since it does not require the work of removing the bracts from the buds in the laboratory and their drying to obtain the dry weight. These authors also proposed in their work with apricots, that the dormancy of the buds had already been overcome when the buds present a minimum increase of 30% in the fresh weight. Both methods, with whole buds or only clean floral primordium provided accurate results (Andreini et al. 2014).

Although the test of Tabuenca was created to evaluate species of the genus *Prunus*, which have the same pattern of flower bud formation, this method was also used to study the dynamics of apple bud dormancy in temperate climate conditions in France and in the mild winter in Brazil, demonstrating also to be suitable for this species (Malagi et al. 2015). The mixed morphology of the floral bud of apple tree allowed to obtain results similar to those found by the test of isolated cuttings with the vegetative buds.

According to these authors, the end of endodormancy found by the dry weight analysis of the floral buds showed a strong correlation with the end of the transition phase of the endodormancy for the vegetative buds ecodormancy, being an efficient method in both temperate and mild winter conditions. In this research, branches with floral terminal buds were used, which were subjected to forcing for 7 days in an air-conditioned room at 24°C and 16 hours of photoperiod with the base immersed in water. The floral buds were peeled (bud scales removed) to obtain the fresh and dry weight of the flower primordia after dehydration at 90°C. Endodormancy ending was defined when the first significant difference between weights of floral primordial weights occurred (Figure 3). Differences were found in determining the endodormancy end date between the methods using the fresh and dry weight of the buds. This was also observed in the study by Yaacoubi et al. (2016) with apple and apricot, and the significant differences between the buds weight before and after the forcing occurred earlier when the fresh weight method was used, mainly with the cultivars of apple trees in France, hastening from 8 to 35 days depending on the year. In Brazil, in mild winter conditions, this difference did not occur, or was lower, ranging from 7 to 14 days.

Regarding the sample size, 50 buds from 3 to 4 branches were used for each sample in the initial tests with peach, apricot (Tabuenca 1964) and plum (Tabuenca 1967). In pear trees, only 10 buds were used, since they have about 6 to 10 floral primordia in each bud (Tabuenca 1964, 1968a). In these initial studies the endodormancy ending was considered when the mean dry weight of the buds increased. It is recommended that a statistical analysis be performed considering the endodormancy when the first significant difference between the weight of the buds without and with forcing occurs.

It is necessary to predict the number of repetitions adequate to the planned experiment and to use samples with a representative and reliable number of buds, which should not be less than 5 buds. It is also recommended that the scales of the buds be removed with the aid of a stereoscopic microscope to avoid damages in the flower primordia.

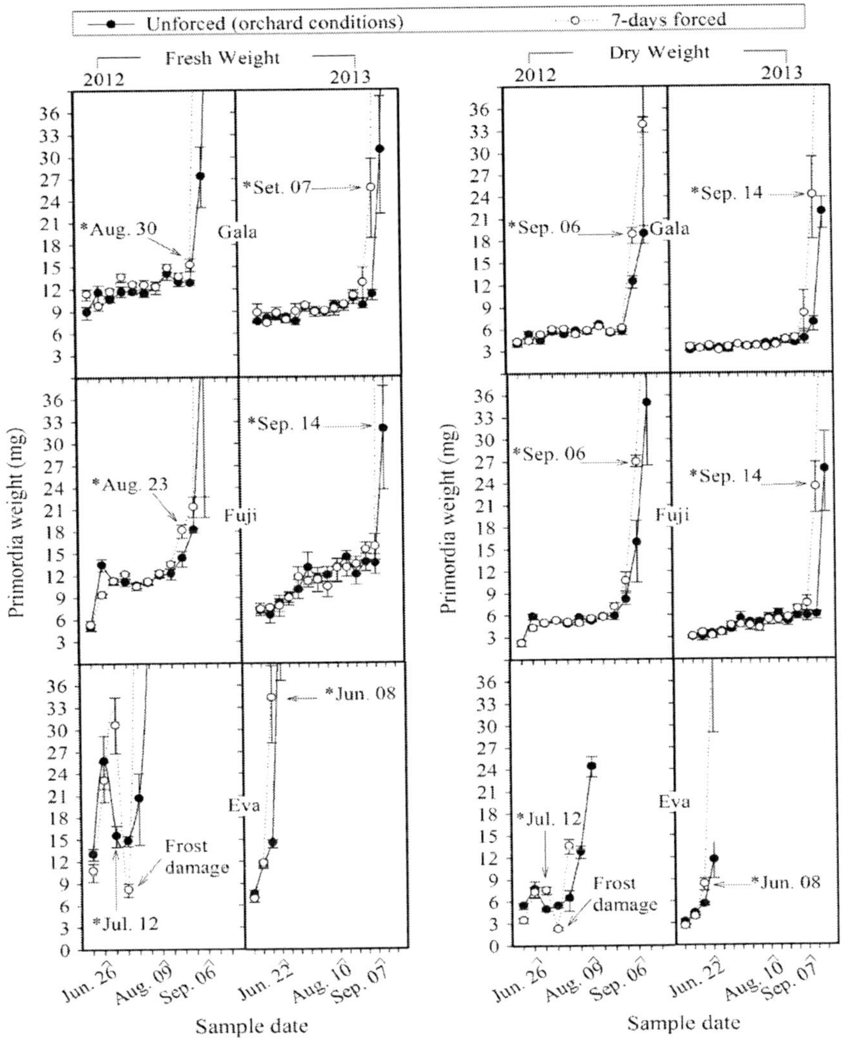

Figure 3. Fresh and dry weight of flower primordia before and after 7 days of forcing, for 'Eva', 'Fuji' and 'Gala' apple cultivars, during 2012 and 2013, Palmas, Paraná State, Brazil. The vertical bars represent the confidence interval at 5% of significance. *The dates followed by an arrow indicate the first date when a significant change ($p < 0.05$) in weight was induced by forcing (Malagi et al. 2015).

3. The Single Node Cutting Test

The single node cutting test is the most used test to study the dormancy of buds of temperate fruit plants. The test was proposed by Pouget (1963) to evaluate the dormancy and its overcoming in grape buds. Cutting test of isolated nodes allows the evaluation of dormancy dynamics and its depth, different from the method of Tabuenca, which only allows to determine when dormancy has been overcome.

The method is simple and involves basically the collection of branches formed in the previous growing season, which are segmented into small cuttings containing only a single bud. The size of the cuttings is usually about 6 to 10cm. For vines and kiwis, which have long internodes, it is recommended a cutting of each node. For species that have short internodes, it is necessary to remove the excess buds, as in the case of apple, peach and plum trees. These buds should be removed preferably with a scalpel so that a smooth and undamaged cut is made, also observing that the meristematic region is completely removed, otherwise sprouting may occur, damaging the test response. The bud that will be evaluated should be left in the upper portion of the cutting, but preserving a distance of at least 1cm from the apical cut. This prevents the loss of the bud by dehydration or a possible fungal infection. This cut can also be protected with paraffin or plastic film.

After the cuttings are prepared, they are brought to an environment with ideal growth conditions, where the temperature and the photoperiod are controlled. The temperature normally used is between 20 to 25°C and the photoperiod of 16 hours of light. Humidity must be kept high to avoid dehydration of the cuttings, which is easily achieved within BOD chambers with automatic humidity control, or into closed plastic containers where the substrate supporting the cuttings is kept moist.

The substrates have only the function of supporting the cuttings, so substrates that are free of mineral nutrients such as phenolic foam and vermiculite are normally used. When plastic boxes are used, they must be transparent to allow the passage of light and also allow the visualization of the buds (Figure 4). It is also possible to leave the cuttings with the base immersed in water using some support so that they can remain upright throughout the test (Malagi et al. 2015). However, in this way the cuttings are at greater risk of clogging of the conducting vessels by the occurrence of fungi and cause the dryness of the buds, harming the sprouting, as verified for apple trees (Anzanello et al. 2014).

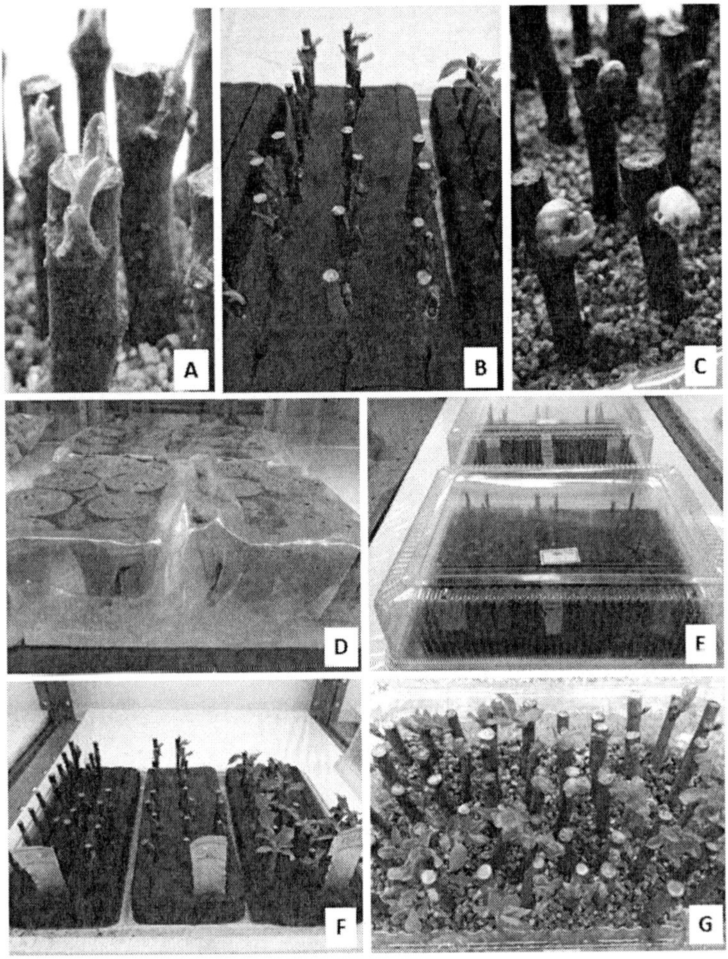

Figura 4. Single nodle cutting test. A. Apple cuttings with open bud. B. Phenolic foam with apple cuttings. C. Grapevine cuttings in vermiculite. D. Cuttings in pots covered with plastic bags. E. Phenolic foam within transparent plastic boxes. F. Apple cuttings with buds at different stages of development. G. Blueberry cuttings in boxes with vermiculite. Source: Biasi.

Because the environment is favorable to the occurrence of fungi and the relatively long time of the test, it is recommended to disinfect the cuttings with germicidal agents such as sodium hypochlorite or to treat the cuttings with fungicides before starting the test. It may also be necessary to re-spray the cuttings during the test if the occurrence of fungi is detected. The high temperature and humidity of the environment allow a very fast growth of the fungi, which can cause the death of the buds in few days.

Test time may vary depending on the species or type of study, but it usually lasts between 30 and 40 days. In a study with apple buds, sprouting took place up to 35 days after the test, with only 0.61% of the buds sprouting after this period (Anzanello et al. 2014). However, the time for sprouting may vary with the chilling requirement of the species and the date of collection, as observed in 'Bruno' kiwifruit whose average sprouting time was higher than 40 days during the period of deeper endodormancy, whereas for the grapevine 'Niagara Branca' the mean sprouting time was less than 25 days (Biasi et al. 2012).

The most important variables obtained by the test are the budburst rate (BR) and the average budding time. The sprouting rate is obtained by the percentage of buds that sprouted in relation to the number of buds at the beginning of the test.

$$\mathrm{BR} = \frac{Nb\ x\ 100}{Ni}\ (\%)$$

Nb = number of budburst

Ni = number of initial buds

The buds are considered sprouted when they reach the green tip (GT) stage. For each species, this stage may have particular characteristics, so it is important to know the visual aspect of the buds during their development, in order to carry out a more precise evaluation. The characterization of shoot stages of the plum tree, persimmon, apple, pear, peach, kiwis and vine shoots was carried out by Carvalho et al. (2010a) in order to standardize and facilitate test assessments.

The mean time of budburst (MTB) is calculated as the mean number of days between the start of the experiment and the day that GT stage was observed for each bud.

$$\text{MTB} = \frac{\Sigma Di}{Nb} \text{ (days)}$$

D = days for each bud reached the GT stage

Nb = number of budburst

With the test it is also possible to calculate the velocity of budburst (VB) of the buds, which is a result variable opposite to the average budding time, since whenever the average time for sprouting the buds is higher, the budding speed will be lower (Figure 5). The velocity of budburst (VB) is calculated using the following equation:

$$\text{VB} = \Sigma \frac{Ni}{Ti} \text{ (buds/day)}$$

N_i = number of buds that reached the GT stage at time "i"

T_i = days passed from the start of the experiment

In the single node cutting test, the great majority of authors have always considered the green tip stage as the end of the budding evaluation. However, although the green tip stage confirms the return of cell division activity in the meristematic region of the bud, it does not necessarily mean that this bud will evolve to a more advanced stage. One of the factors that can affect sprout growth is the amount of reserve available at the cutting. Therefore, it is important to use cuttings that have a diameter considered of good vigor. Considering that the cuttings have sufficient reserves to evolve from the green tip stage, it is interesting to continue the evaluation of the test for a longer time until the buds reach a more advanced stage, such as open bud (OB), when the first leaves become visible. In this way, the vigorous bud rate (VBR) can be calculated by the percentage of cuttings with buds reaching the open-bud (OB) stage in relation to the cuttings that had reached the green tip (GT) stage.

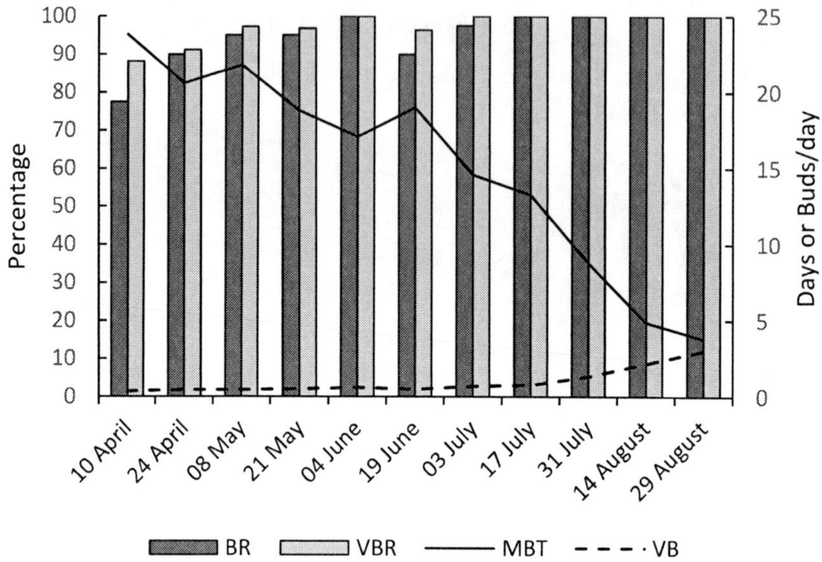

Figure 5. Mean budburst time (MBT), budburst rate (BR), velocity of budburst (VB) and vigorous bud rate (VBR) in the single-node cutting test performed with 'Niagara Branca' grapevine collected in Colombo, Paraná State, Brazil during the autumn and winter in 2009.

$$\text{VBR} = \frac{Nob\ x\ 100}{Ngt}\ (\%)$$

Nob = number of buds that reached the OB stage

Ngt = number of buds that reached the GT stage

In a study conducted in a region with mild winter, with only 367 chilling units calculated by the Utah method, the rate of vigorous shoots was close to 100% for the 'Imperial Gala' apple buds throughout the evaluated autumn and winter period. However, for the 'Hosui' pear buds this rate ranged from 18.9 to 67.9%, reaching 95% when at the end of the test the branches received 1440 hours of additional chilling. These results demonstrate the importance of continuing to evaluate budding beyond to the green tip stage, especially in more demanding chilling species, which, in addition to presenting a low budding rate at the green tip stage, few of these buds evolve to the open ones (Carvalho et al. 2012). A fact also observed in the evaluation of the dormancy dynamics during the autumn and winter period in kiwis and vines, where the rate of vigorous shoots was close to 100% for the grapevine, but much lower for the kiwi (Biasi et al. 2010).

The single nodes test allows the evaluation of the endodormancy of the buds, eliminating the paradormancy effect that can occur between the buds. Therefore, a higher sprouting rate is observed in the lateral buds with the isolated node cuttings when compared to the use of whole cuttings. In a study carried out with apple trees, sprouting of lateral buds was 83% in cuttings with only one bud, while with whole cuttings the sprouting of this type of bud was only 11.3%. This result is attributed to the presence of apical buds that exert an apical dominance effect. However, the apical buds are free of this paradormancy effect, because when compared in cuttings with only one bud and whole cuttings, they presented 67.3% and 73.2% sprouting, respectively (Anzanello et al. 2014). Even with the removal of the apical bud on whole cuttings, the superior bud can assume the effect of apical dominance, inhibiting sprouting of the lateral buds, especially in cultivars that are more demanding in chilling for dormancy release (Putti et al. 2003).

Isolated node cuttings are more susceptible to losses due to the greater exposure of the internal tissues to contaminations and to the effect of the action of ethylene and abscisic acid induced by cuts in the tissues close to the bud (Dennis Junior 2003).

Methods using whole plants or grafted branches to evaluate bud dormancy prevent branch losses from contamination or dehydration and can be evaluated for a longer period than detached cuttings. However, they are impractical because they require a very great space for their accomplishment (Camelatto et al. 2000; Wagner Junior et al. 2006; Chavarria et al. 2009; Hawerroth et al. 2010).

4. Dormancy Index

The buds are formed by several tissues with distinct composition and purpose and thus their physiological state of dormancy is the result of the individual and particular metabolism that occurs in each cell that constitutes it. In one branch there may be buds with deep dormancy and others with dormancy overcome. Among the buds with established dormancy, there are the most dormant and the least ones. The same bud may have in its composition leaf primordia suitable for cell division and elongation and still dormant meristematic tissues. These characteristics make it very difficult to evaluate the dormancy of a bud, especially in biological tests (Figure 6).

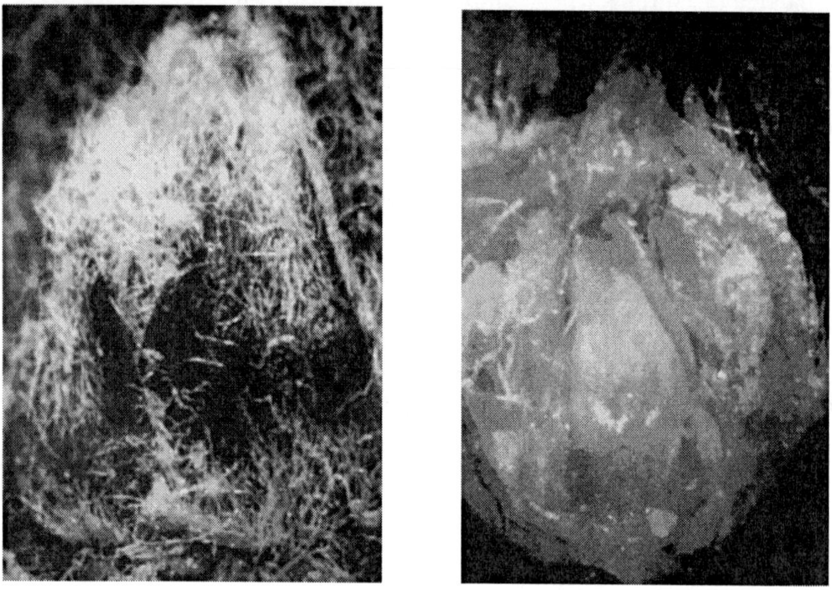

Figure 6. External morphology (left) and internal (right) morphology of one-year-old buds of apple trees (Carvalho and Zanette, 2005a).

In the biological test of single node cuttings, it is expected that a bud with the dormancy completely overcome will start sprouting and continue the process of elongation of the first leaves until the bud is fully open in short time. A bud with dormancy partially overcome in its cells or tissues may present sprouting, without continuing the process of cell elongation, since the internal physiology of the bud is not totally directed to growth in a regular and constant way.

In order to conclude more precisely about the dormancy release in the biological test, the buds should be observed for a longer time in order to make sure that the whole set of tissues was able to the growth and the formation of an open bud. It is not unusual for a group of buds to initiate their opening by reaching the green tip (GT) stage, but does not evolve until the first leaves expand. Thus, in addition to the determination of the mean time of budburst (MTB), which is the time elapsed between the installation of the test and the detection of the GT stage, other variables must be observed to determine if the bud had its dormancy overcome or not and in which intensity of dormancy. The faster a group of buds starts the sprouting process and in a more synchronized way, there is a higher velocity of budburst (VB) which is a strong indication that dormancy has been completely overcome.

Determination of the percentage of buds that began sprouting, continued the process of cell stretching and expanded first leaves characterizes this bud ability to sprout vigorously and form new branches. Two variables demonstrate this ability that is the budburst rate (BR) and the vigorous bud rate (VBR). In a biological test, it is possible to occur that a group of buds quickly initiated the budding process forming the green tip and also rapidly evolved into the open-bud stage. These characteristics are expected for buds whose dormancy has been completely overcome. However, there may also be a group of buds that rapidly reach the green-tipped stage, but do not evolve to the open bud, indicating that other physiological events required for the continuity of sprouting have not occurred or are not synchronized. Thus, two groups of buds must have equal average sprouting time, but with a different final sprout rate, which means that one group had its dormancy totally overcome and another one did not.

The same happens with the other variables to be analyzed in the biological test and, only with a joint analysis of all of them and considering their relations, it is possible to reach at a more precise and more comprehensive conclusion on the physiological state of the bud dormancy. This was the basis for the development of the Dormance Index (DI) developed by Carvalho and Silva (2010).

The general formula of the dormancy index was initially determined by the results obtained in the biological test of 'Gala' apple genotypes collected from April to August (autumn and winter) in an area of low chilling accumulation in the South Brazil (latitude 25.55° South, longitude 49.90° West and altitude of 795 m).

Table 1. Correlation coefficients between the mean sprouting time (MTB), budburst rate (BR), velocity of budburst (VB) and vigorous bud rate (VBR) in the biological test of dormancy of apple trees

Parameters	MTB	BR	VB	VBR
MTB	-	-0.9413	-0.9537	-0.8180
BR	-0.9413	-	0.9748	0.7460
VB	-0.9537	0.9748	-	0.6726
VBR	-0.8180	0.7460	0.6726	-

Source: Carvalho and Silva (2010).

The correlation analysis between the four variables analyzed in the biological test (MTB, BR, VB and VBR) shows that there is a distinct relation between them. MTB has a significant negative correlation with the other variables. The lowest correlations were found when one of the variables compared was VBR (Table 1).

Normally, the lower the mean time of budburst (MTB), the lower is the dormancy intensity, in other words, they are directly proportional, but the higher the budburst rate (BR), vigorous bud rate (VBR) and velocity of budburst budding (VB), the lower the intensity of dormancy, that is, they are inversely proportional. Thus, the original basic formula had the following structure:

$$DI = \frac{MTB}{BR+VB+VBR}$$

The sum of the variables BR, VB and VBR was adopted in the denominator of the formula, since it is common for BR or VBR to be equal to zero in the months of intense dormancy in the biological test, which makes it impossible to use the multiplication among the variables for the calculation of the Dormancy Index. As BR and VB were considered the variables with the highest correlation with MTB in the interpretation of dormancy intensity, two multiplication constants were added to them. The constants "k" and "w" were calculated for the variables BR and VB, respectively, by means of the result of the correlation (r) between them. The formulas used to calculate the constants "k" and "w" are those quoted below:

$$|k| = |r_{(BR/VB)}| \times |r_{(BR/VBR)}| / |r_{(BR/VB)}| - |r_{(BR/VBR)}|$$
$$|w| = |r_{(VB/BR)}| \times |r_{(VB/VBR)}| / |r_{(VB/BR)}| - |r_{(VB/VBR)}|$$

$|r_{(x/y)}|$ are the modulus of the correlation coefficient between the variables "x" and "y".

Analyzing the whole data set obtained in the biological test with apple buds, the constants k = 3 and w = 2 were obtained. The general formula of the Dormancy Index of apple trees was determined:

$$DI = \frac{MTB}{3BR+2VB+VBR}$$

After determination of the formula, a spreadsheet with all the possibilities of answer of the buds in a biological test was set up and the values obtained were grouped and classified into five groups of dormancy intensity:

- Absent dormancy;
- Weak dormancy;
- Moderate dormancy;
- Intense dormancy;
- Deep dormancy.

The threshold values of each group are shown in Table 2.

It can be exemplified that in a group of buds in which the MTB found was less than 10 days and the final rates of sprouting and vigorous shoots were higher than 80%, dormancy has already been overcome. In this same case, if the MTB is less than 10 days, but the BR is only 20%, the DI will indicate a moderate dormancy, since many buds may have reached the Green Tip, but were not with their metabolism perfectly balanced for continuity of the budding process.

Table 2. Classification of dormancy intensity according to dormancy indexes obtained for apple buds

Dormancy Index Group (DI)	Dormancy Intensity
DI ≤ 2	Absent dormancy
2 < DI ≤ 5	Weak Dormancy
5 < DI ≤ 9	Moderate dormancy
9 < DI ≤ 14	Intense dormancy
DI > 14	Deep dormancy

Source: Carvalho and Silva (2010).

After the determination of the final DI formula, the data collected in the biological test obtained by Carvalho and Zanette (2004) were applied in the formula and the results found could be interpreted more clearly by means of the analysis of the Dormancy Index only. Table 3 shows the mean data of the variables that compose the DI and the final values found. It was clear that the dormancy started with low intensity in the month of April, increasing intensity in the advance of the autumn and beginning of winter, and the dormancy release at the end of August, after the buds treatment to dormancy release, reducing the DI to values below 2.0.

It is important to note that for the application of the BR and VBR data that are obtained in percentage in the formula, the absolute values are used, which means that for 100% of sprouting, BR = 1.0 or 70% of vigorous sprouting is used the value of VBR = 0.7.

Therefore, all the experiments conducted to evaluate the buds dormancy by the biological test can be evaluated more comprehensively and accurately through the Dormancy Index, facilitating the global analyzes and the comparison between treatments. In the conventional analyzes of the MTB alone for comparison of varieties and or products to overcome dormancy, it can be concluded that there are differences between the treatments, but without the need to determine if any of them were really efficient for the exit of dormancy in an integral way.

Table 3. Classification of dormancy intensity by Dormancy Index calculated by means of mean budbrust time (MTB), final budburst rate (BR), budburst velocity (VB) and vigorous budburst rate (VBR) obtained by Carvalho and Zanette (2004)

Dates	MTB (days)	BR	VB (buds/day)	VBR	Dormancy Index	Dormancy Intensity
04/19	14.0	0.95	0.71	0.92	2.70	Weak
05/10	21.7	0.20	0.11	0.75	13.82	Intense
05/31	21.2	0.55	0.28	0.96	6.69	Moderate
06/21	23.4	0.18	0.08	0.67	17.08	Deep
07/12	28.5	0.08	0.03	0.00	95.00	Deep
08/02	22.9	0.18	0.08	0.50	19.08	Deep
08/23	11.2	0.98	0.92	1.00	1.94	Absent

Table 4. Formulas for calculating the Dormant Index of plum, persimmon, peach, kiwi and grapevine buds

Species	Value of constants		Final formula
	K	W	
Plum	5	3	ID = MTB.(5.BR + 3.VB + VBR)$^{-1}$
Persimmon	4	1	ID = MTB.(4.BR + VB + VBR)$^{-1}$
Peach	3	4	ID = MTB.(3.BR + 4.VB + VBR)$^{-1}$
Kiwi	8	1	ID = MTB.(8.BR + VB + VBR)$^{-1}$
Grapevine	3	2	ID = MTB.(3.BR + 2.VB + VBR)$^{-1}$

Source: Carvalho and Biasi (2012).

Due to the adequate application of the DI for apple buds, Carvalho and Biasi (2012) continued the research and carried out a similar study to determine DI for other temperate fruit species such as plum, persimmon, peach, kiwi, and grapevine.

The biological tests were carried out in the months of April to September of the years 2007 and 2008 with buds collected in adult orchards located in areas of low chill occurrence. It was studied plum buds cv. Poly Rosa, persimmon cv. Fuyu and peach cv. Chimarrita collected in the municipality of Fazenda Rio Grande, Paraná State, Brazil (25° 39' S, 49° 16' O and 895 m a.s.l.) and grapevine buds cv. Niagara Branca and kiwi buds cv. Bruno collected in the municipality of Pinhais, Paraná State, Brazil (25° 23,303' S, 49° 07,993' O and 930 m a.s.l.).

After the detailed analysis of the data in the same way as carried out by Carvalho and Silva (2010), the specific formulas for calculating the DI for each species were obtained (Table 4).

For most of the species, in addition to the average time for sprouting, the final budbrust rate (BR) is the most important variable for obtaining DI. Only for the peach trees the sprouting speed had a greater influence than the BR in the DI calculation. The dormancy intensity groups were also determined and the values are described in Table 5.

It is concluded that with the specific DI formulas and the defined interpretation tables, all the experiments that involve the biological test for these species can be evaluated through DI in a more precise and applied field. Once the method for determining the formulas has been determined, other studies with other fruit species can be carried out increasing the range and application of the Dormancy Index.

Table 5. Classification of dormancy intensity according to the Dormancy Indexes obtained for plum, persimmon, peach, kiwi, and grapevine buds

Dormancy	Species				
Intensity	Plum	Persimmon	Peach	Kiwi	Grapevine
Absent	0 to 0.6	0 to 2.5	0 to 0.6	0 to 1.2	0 to 1.0
Weak	0.61 to 1.8	2.51 to 4.5	0.61 to 1.8	1.21 to2.2	1.1 to 2.5
Moderate	1.81 to 3.5	4.51 to 8.0	1.81 to 3.5	2.21 to 6.0	2.51 to 5.5
Intense	3.51 to 8,0	8.1 to 12.0	3.51 to 8.0	6.1 to 9.0	5.51 to 8.0
Deep	>8.0	>12.0	>8.0	>9.0	>8.0

Source: Carvalho and Biasi (2012).

5. Biochemical Tests

Biochemical tests for dormancy are based on variation of ATP content (Bonhomme et al. 2000), protein content (Tamura et al. 1998), carbohydrate dynamics (Carvalho and Zanette 2005b), the fatty acid content in the phospholipids of the buds cell membranes (Erez 2000), the amount and activity of the ATPase in the plasma membrane of the adjacent buds and tissues (Aue et al. 1998; Aue et al. 2000) and the intensity of bud breathing (McPherson et al. 1997).

The ability to synthesize adenosine triphosphate (ATP) and non-adenine nucleotides (NTP) in vegetative buds can be evaluated after the incubation of the buds in adenosine and non-adenosine medium, precursor of these nucleotides. The increase in the level of NTP and ATP reveals the ability of the bud to synthesize and the absence of endodormancy without considering the preliminary environmental conditions (Bonhomme et al. 2000). The level of ATP is low during the rest period and a large increase in the ATP/ADP ratio (phosphorylation degree index) occurs in flower and vegetative buds after a long period of exposure to cold. This high ratio may be indicative of the bud potential to establish a new sprout with vigorous growth. The measurement of these contents and their relative indexes can represent a way of evaluating dormancy "in situ", without considering the preliminary environmental conditions (Bonhomme et al. 1999b; Bonhomme et al. 2000). This method, however, is not suitable for flower buds (Bonhomme et al. 1999a).

The carbohydrate content test is based on the hypothesis that the growth control of the buds during the winter is due to a change in the nutrient flow to the tissues very close or adjacent to the meristematic zone, causing a decrease in the supply of these nutrients and, consequently, blocking the meristem growth. The main carbohydrates analyzed are starch, sucrose, glucose, fructose and sorbitol (Bonhomme et al. 1999c). The alteration of the carbohydrate level allows the elaboration of curves that show its dynamics and thus determine the growth potential of the buds.

Water movement can have important physiological consequences because, together with water, soluble reserves (sucrose, sorbitol) of the bud can be exported to adjacent tissues, increasing the content of insoluble reserves (starch) in the bud. This imbalance may lead to a future blockage to the nutrition of leaf or flower primordia when the bud meets sprouting stimuli. However, water accumulation in the bud may not be indicative of resuming growth because water may be unavailable in the tissue, not initiating the most important biochemical reactions to the metabolism of the bud (Erez et al. 1998).

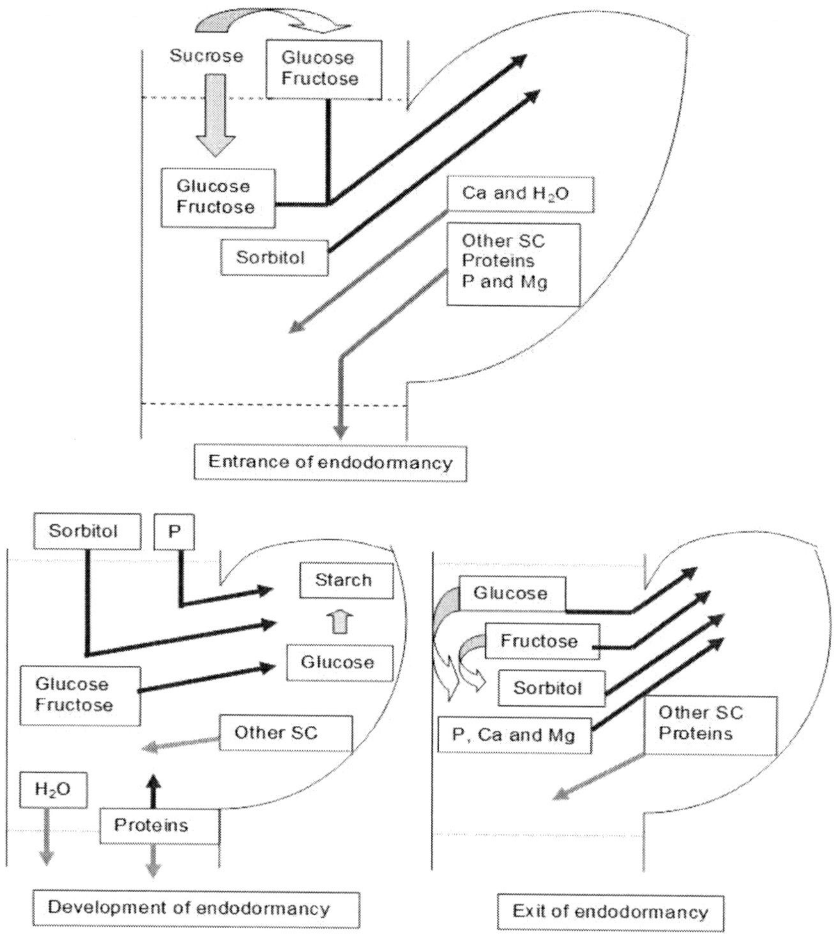

Figure 7. Dynamics of water, carbohydrates, proteins and minerals during the entrance, development and exit of the endodormancy of 'Imperial Gala'. (Carvalho, 2001).

The higher percentage of soluble carbohydrates in relation to the insoluble ones at dormancy suggests a greater predisposition to sprout, while the higher proportion of insoluble carbohydrates in the more intense dormancy suggests predisposition to storage for future use. With the entry into the dormant period, the reduction of the soluble carbohydrate content in the buds also occurred in the branches, indicating that in this stage, there is also transport of carbohydrates to regions of the branch farthest from the buds, even if the migration of water has not occurred at this stage (Figure 7). Soluble and insoluble carbohydrates represented 13.8 to 20.2% of the dry matter of one-year bud and 9.9 to 15.3% of the dry matter of the one-year branches of 'Imperial Gala' (Carvalho and Zanette 2004).

The type of transport in these tissues occurs at very short distances and can occur via symplast or apoplast (Thorpe et al. 1996). Transport via symplast, by plasmodesma, can exert a strong control of molecules selectivity, facilitating the small molecules passage and blocking the passage of larger ones. The monosaccharides glucose, fructose and sorbitol are smaller molecules than the other oligosaccharides, so they could be transported via plasmodesmas in any direction. Transport via apoplast involves membrane carriers (proteins) and can occur along with the transport of other substances in the same direction (simporte) or in the opposite direction (antiporte) (Buckhout and Tubbe 1996).

Respiratory and oxidative stress has been related to the bud dormancy process in several species. The stress caused by the low temperatures and the decrease of the respiratory activity during the rest period induces an increase in the production of reactive oxygen species (ROS), such as superoxide (O^{-2}) and hydrogen peroxide (H_2O_2). The H_2O_2 has a relatively long life and can pass freely through the membranes by diffusion, acting as a messenger in stress signaling. ROS at high concentrations may be toxic to cells and cause damage such as lipid peroxidation of the membrane, but may also be the signs of overcoming dormancy. ROS are produced continuously in the cells as a result of respiration in all organelles. Cells have the ability to both produce and rapidly eliminate the different forms of ROS, as a result of a balance between formation and detoxification, avoiding the damages resulting from their accumulation. This characteristic makes the ROS, signaling substances of the changes in the cellular metabolism (Pérez and Lira 2005; Mittler et al. 2011; Vergara et al. 2012; Beauvieux et al. 2018).

ROS are produced mainly in mitochondria, chloroplasts and peroxisomes (Bowler et al. 1992; Pérez and Lira 2005; Mittler 2017).

Mitochondrial respiration is an important source of ROS under normal conditions and this production is elevated in abnormal conditions, such as hypoxia during dormancy (Vergara et al. 2012; Beauvieux et al. 2018; Meitha et al. 2018). Reactive nitrogen species (RNS) are also associated with oxidative stress, such as nitric oxide (NO), a gaseous substance that can serve as a potentiator in the ROS generation. The rapid accumulation of sublethal amounts of ROS and RNS induced the loosening of the cell wall and expansion for sprouting of dormant buds (Sudawan et al. 2016).

The main antioxidant defense mechanisms in plants include the enzymes catalase (CAT), glutathione peroxidase (GR), superoxide dismutase (SOD), guaiacol peroxidase (POD), ascorbate peroxidase (APX) and peroxidase superfamily proteins (Mittler 2002; Viti et al. 2012; Hussain et al. 2015; Takemura et al. 2015; Beauvieux et al. 2018; Farokhzad et al. 2018).

CAT is found in peroxisomes and glyoxysomes (Mittler 2002) and is responsible for the breakdown of the H_2O_2 molecule into water and oxygen (Or et al. 2002; Pérez and Lira 2005). SOD can be found in apoplast, chloroplast, cytosol, mitochondria and peroxisome and is responsible for catalyzing the dismutation of superoxide into oxygen in hydrogen peroxide (Wang et al. 1991; Mittler 2002). APX is located in the apoplast, chloroplast, cytosol, mitochondria and peroxisome (Mittler 2002), where it detoxifies H_2O_2 using ascorbate as a substrate that donates electron to H_2O_2, producing dehydroascorbate and water (Ahmad et al. 2010). POD is found in the cell wall, cytosol and vacuoles. This enzyme uses guaiacol as a substrate to donate electrons to H_2O_2 and form oxidized ascorbate and water (Mittler 2002; Asada 2006).

In grapevines, it was verified that the control of the dormancy of buds is under the effect of ROS and antioxidant enzymes. Catalase activity decreases during exposure to low temperatures during the winter and causes an increase in the levels of hydrogen peroxide in the buds, which initiates a process of signal transduction, which results in sprouting buds when climatic conditions are favorable (Nir et al. 1986; Pérez and Lira 2005; Sudawan et al. 2016, Khalil-Ur-Rehman et al. 2017).

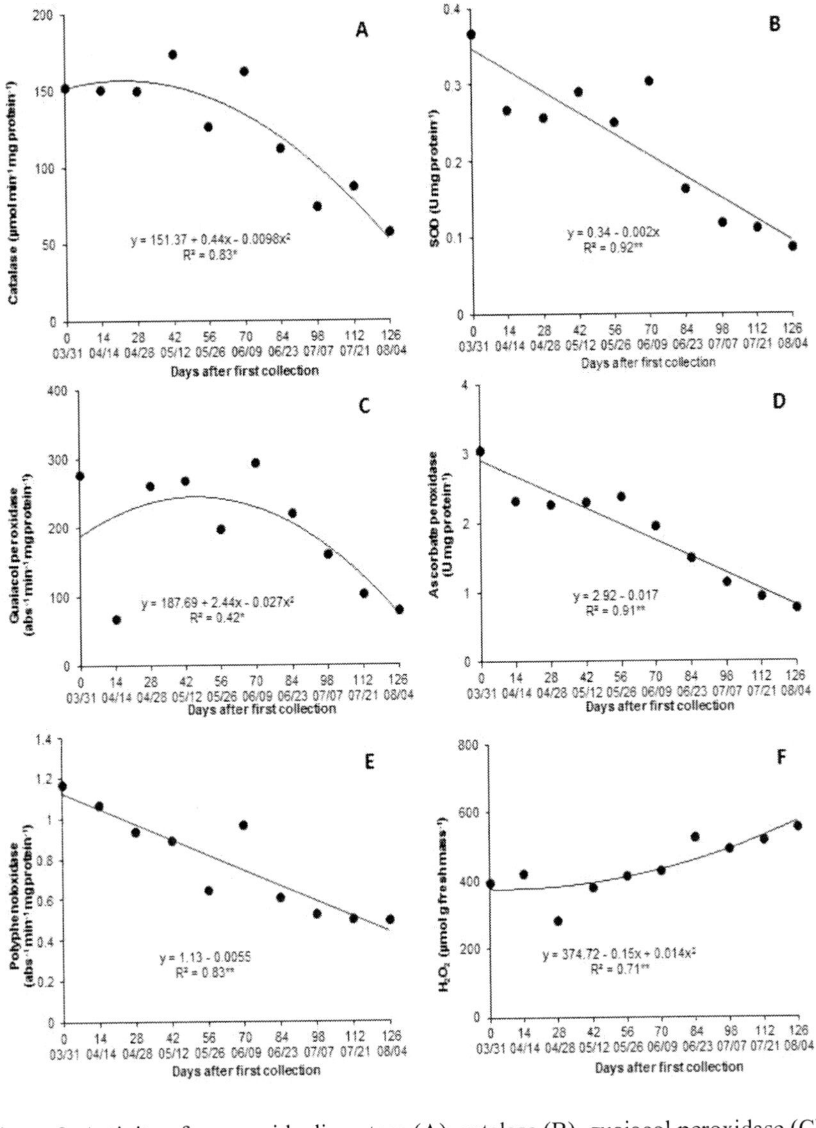

Figure 8. Activity of superoxide dismutase (A), catalase (B), guaiacol peroxidase (C), ascorbate peroxidase (D) and polyphenol oxidase (E) and hydrogen peroxide content (F) of persimmon 'Fuyu' during the period from March to August 2015 in Campina Grande do Sul, PR, Brazil (Pereira, 2017).

The oxidative stress was studied in buds of the 'Fuyu' persimmon, where it was observed that the activity of the CAT, SOD, POD and APX enzymes decreased over time and with the accumulation of cold hours when the buds were induced to enter endodormancy (Figure 8). The reduction in the activity of the antioxidant enzymes was accompanied by the increase in the hydrogen peroxide content until the release of the buds dormancy, which occurred when the H_2O_2 contents were high, suggesting the possible involvement of this ROS in the release of the persimmon buds dormancy. The PPO activity also decreased, suggesting an increase in phenolic compound contents, which may have contributed to the ROS not reaching toxic levels and causing bud damage. Therefore, the increase of the H_2O_2 contents must have been maintained in levels favorable to overcoming dormancy without reaching toxic levels (Pereira 2017).

The analysis of the levels of antioxidant enzymes and ROS, mainly of H_2O_2, can be used to evaluate the oxidative stress and to follow the evolution of bud dormancy. Therefore, studies are necessary to verify this metabolism in temperate fruit species and to establish the levels of these substances to correlate with the periods of entry, permanence and exit of endodormancy.

6. Tetrazolium Test

The methods used to assess the dormancy of temperate fruit buds may be highly correlated with actual dormancy intensity, but they are laborious, time-consuming or require specific resources that make it unusable for large-scale use. Thus, already developed methods and theories regarding the bud endodormancy contribute much to the science and specific studies, but they have restricted applicability and do not allow to the producer the dormant answer of the bud in a short period, less than 24 hours, to support your decision-making on what to accomplish in your orchard. In this context, a new method of bud dormancy analysis was developed by Carvalho et al. (2010b) which it is efficient, fast and simple. The basis of the new method is the respiratory intensity of the bud tissues.

Cellular respiration in plants consists of the breakdown of molecules to obtain energy, in the form of adenosine triphosphate (ATP), for the growth of cells and tissues. As in plants these molecules come from photosynthesis, which by means of the transduction of light energy and carbon fixation synthesizes the necessary carbohydrates, respiration represents part of the flow of energy in the plant. Glucose is the main end product of the hydrolysis of sucrose and starch and its oxidation is the beginning of the respiratory process in cells. In oxidation, the glucose molecule is cleaved and the electrons and protons are removed from the carbon atoms and combined with the oxygen, which is thereby reduced. Electrons go from higher levels to lower energy levels, releasing it (Raven et al. 2001).

During the bud dormancy period, although the plant does not show visible growth, essential metabolic activities, such as cellular respiration, continue to occur, however, with reduced intensity (Petri et al. 1996). In this stage, energetic metabolism changes in the bud, such as enzyme activity and nucleotide synthesis, which may be involved with the dormancy intensity and, consequently, the budding ability (Balandier et al. 1993; Bonhomme et al. 2000). The return of metabolic activity at higher intensity may be an indication of preparation for budding, indicating that dormancy has been exceeded.

Dehydrogenase enzymes are involved with respiratory activity in biological systems and during the respiratory processes, intermediates are formed and serve as substrates for these enzymes that release hydrogen ions (Dias and Barros, 1995). Thus, respiratory intensity is associated with greater hydrogen ions release.

The tetrazolium test is a method used to estimate the viability of seeds in a few hours and is based on the oxidizing and reducing characteristic of the tetrazolium salt (2,3,5-triphenyl tetrazolium chloride), which acts as a hydrogen ion receptor released during the process of cellular respiration. Tetrazolium is reduced to a red, insoluble and stable substance, called formazan. As this reduction takes place inside the cells and the pigment is not diffusible, the living tissue acquires red coloration and the dead tissues remain without staining, since there is no breathing. The red staining intensity of the tissues is related to the greater permeability of the same to the salt that thus, can receive more ions hydrogen. In seed, the color intensity is not enough to diagnose the quality of the seed, since it is a vital embryo axis to the formation of a new and viable new plant, it must be totally healthy and vigorous. For this reason, judicious visual analysis is indicated for evaluation of the test (Dias and Barros 1995). For most seeds, the concentration of the tetrazolium solution ranges from 0.1 to 1.0% and the treatment time from 2 to 24 hours depending on the permeability of the tissue to the salt (Brasil 2009).

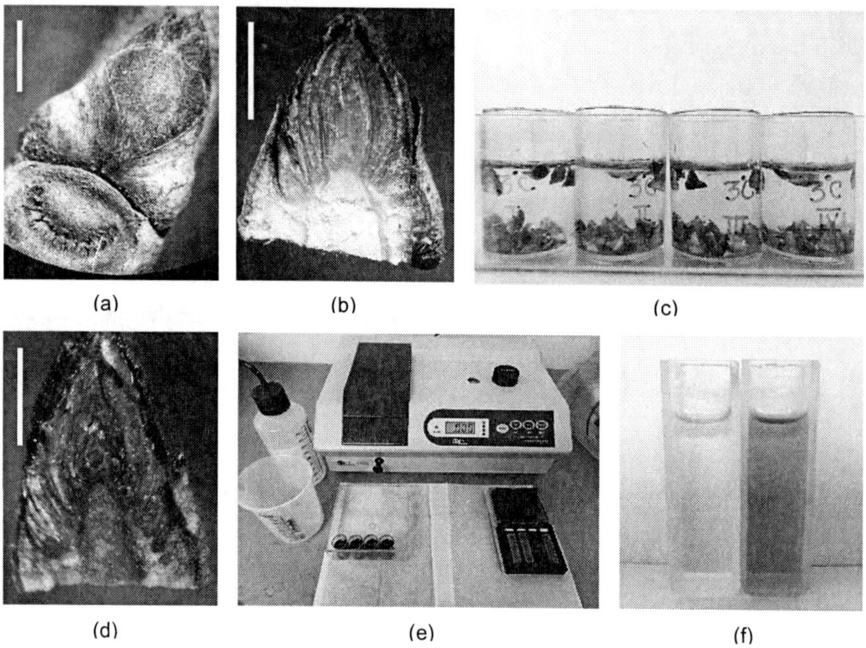

Figure 9. Stages of the tetrazolium test on persimmon buds. (a) Whole bud; (b) longitudinally sectioned bud; (c) Sample of buds in the tetrazolium solution; (d) Colored bud after 2 h immersion in the solution; (e) spectrophotometric reading; (f) Samples of alcoholic solution for reading, control (left) and color treatment (right) (Pereira, 2017).

The method developed by Carvalho et al. (2010b) is based on the premise that the inner living tissues of the bud have respiratory activity and react with the tetrazolium salt with greater or lesser intensity. The greater the tissue dormancy, the less its respiratory activity and the less its coloration in contact with the tetrazolium salt.

The method consists in collecting branches with one year-old buds in the orchard and taking them immediately to the analysis site. The isolated lateral or group buds, as for the peach and the plum, are detached from the branch with a blade with a longitudinal cut between the bud and the branch and another transverse at the point of insertion of the same in the branch. The initial studies were carried out with different sample masses of longitudinally sectioned buds and exposed for different times (2 to 10 h) in 5 mL of solutions with different concentrations of tetrazolium salt (2 to 10 g L^{-1}) for tissue staining at a controlled temperature of 25°C. Different masses of samples were studied in order to obtain a minimum number of buds that is more representative. For species with larger masses, such as kiwi, the samples are larger, and for smaller bud species, such as the plum tree, the samples are smaller. Then the buds are removed from the tetrazolium solution in absolute ethyl alcohol at room temperature for one hour to extract the red coloration of the buds. After this time, the intensity of alcohol staining is done by spectrophotometry by absorbance at 560 nm (Figure 9).

The pear buds presented homogeneous staining at the different tetrazolium salt concentrations, regardless of the treatment time. It is recommended the collection of lateral buds of the 5th to 15th nodes of one year-old branches counted from their apex. The other species have a particular behavior (Figure 10).

The apple buds show more intense color as the concentration of the tetrazolium solution increases, so a concentration of 10 g L^{-1} for the test is recommended for 2 h. It is recommended the collection of lateral buds of the 5th to 15th nodes of branches of one year-old counted from its apex.

In plum buds, the most appropriate red coloration of the tissues was obtained with concentrations of 4.0 to 10 g L^{-1} per 2 h and, in this case, the highest concentration is recommended to allow the test to be performed in months of lower metabolic activity. Samples of groups of lateral buds should be collected from 5th to 15th nodes counted from the apex of branches. Each group of buds is formed by three buds being a central vegetative bud and two floral buds lateral.

For the study of peach bud dormancy, the plum-like method is recommended, with groups of three lateral buds located from the 5th to the 15th nodes counted from the apex of mixed branches, the central vegetative bud and floral lateral buds. It is recommended 10 g L^{-1} of the tetrazolium salt for the test for 2 h.

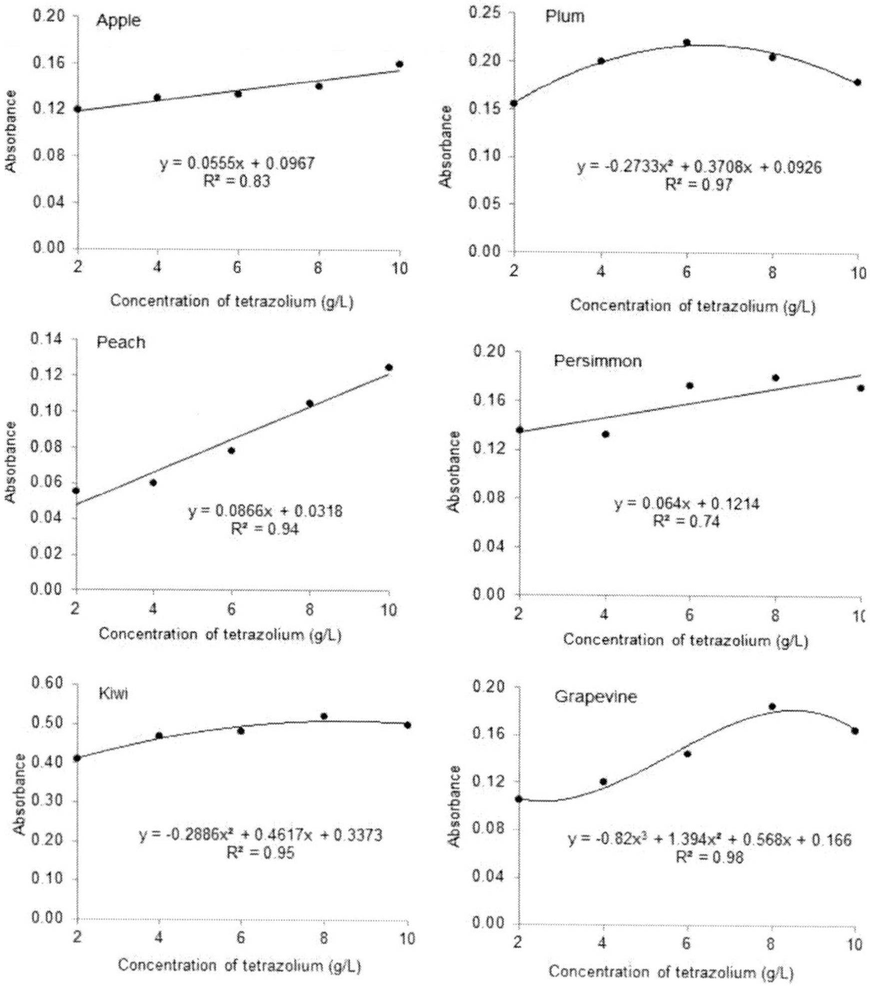

Figure 10. Red coloration absorbance variations of the several species buds after treatment with different concentrations of tetrazolium for two hours (Carvalho et al. 2010b).

The lateral buds of persimmon should be collected from 2nd to 6th nodes counted from the apex of one year-old branches and should be treated with 10 g L^{-1} concentration of the tetrazolium salt for 2 h.

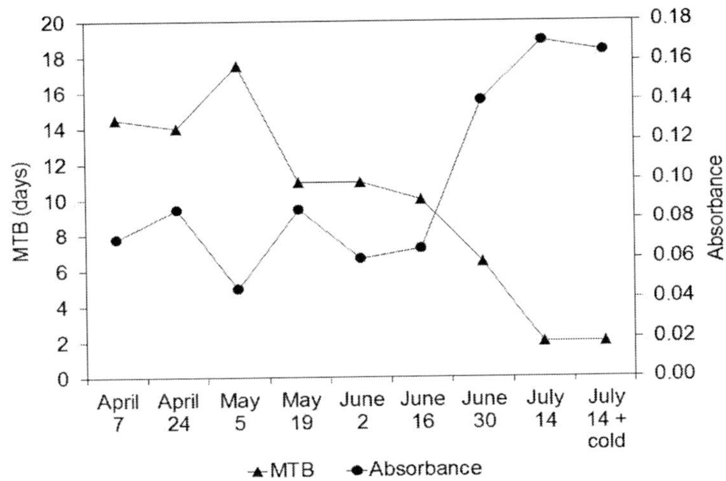

Figure 11. Dynamics of peach bud dormancy obtained by the biological test and the tetrazolium test (r = -0.92).

For the analysis of dormancy of kiwifruit and grapevine, buds from 2nd to 8th nodes counted from the basal portion of one year-old branches should be collected. The buds should be treated with 8 g L^{-1} tetrazolium solution for 2 h.

The correlation between the mean time for budburst (MBT) in the biological test and the intensity of bud staining in the tetrazolium test is inversely proportional, that is, when the MTB is higher the bud staining is lower due to the more intense dormancy. In this way the dormancy intensity can be evaluated much quickly than with the biological test. The peach presented the most significant correlation, indicating that the tetrazolium test is an excellent indicator of the dormancy of their buds (Figure 11).

Table 6 shows the specifications for the accomplishment of the tetrazolium test with several species studied. The need to obtain measurable results was considered for the definition of the recommended protocols even when the buds are deeply dormant, which would result in low respiratory activity, low staining with tetrazolium salt and, consequently, low readings by spectrophotometry. Smaller bud masses, lower tetrazolium concentrations and higher volumes of alcohol for extraction may lead to a lack of final absorbance reading, making it impossible to complete the test.

Table 6. Mass of buds, duration of treatment in tetrazolium solution, concentration of tetrazolium salt and volume of alcohol recommended for the analysis of the coloration of buds of temperate fruit trees by spectrophotometry

Species	Mass of the bud sample (g)	Duration of treatment (h)	Concentration of tetrazolium (g L^{-1})	Volume of ethyl alcohol (mL)
Plum	0.15	2	10	6
Persimmon	0.5	2	10	6
Apple	0.2	2	10	4
Pear	0.3	2	10	4
Peach	0.3	2	10	6
Kiwi	1.0	2	8	6
Grapevine	0.2	2	8	4

Source: Carvalho et al. (2010b).

It is concluded that the use of the tetrazolium test to evaluate the metabolic activity of temperate fruit buds provides the best understanding of bud dormancy in a short period of time and allows a more precise dormancy intensity evaluation of the buds in an orchard, making possible more informed decisions on orchard management, especially the need to apply products to overcome dormancy.

The monitoring of the results of the tests during autumn and winter in the orchard will accurately show the entrance of the dormancy of the buds and their natural or not, almost real-time overcoming, becoming an essential tool for the decision making in the handling of the fruit trees.

References

Ahmad, P., Jaleel, C. A., Salem, M. A., Nabi, G. and Sharma, S. (2010). Roles of enzymatic and nonenzymatic antioxidants in plants during abiotic stress. *Critical Reviews in Biotechnology,* 30(3): 161-175.

Andreini, L., Cortázar-Atauri, I. G., Chuine, I., Viti, R., Bartolini, S., Ruiz, D., Campoy, J. A., Legave, J. M., Audergon, J. M. and Bertuzzi, P.

(2014). Understanding dormancy release in apricot flower buds (*Prunus armeniaca* L.) using several process-based phenological models. *Agricultural and Forest Meteorology,* 184: 210-219.

Anzanello, R., Fialho, F. B., Santos, H. P., Bergamaschi, H. and Marodin, G.A.B. (2014). Métodos biológicos para avaliar a brotação de gemas em macieira para modelagem da dormência [Biological methods for assessment of budbreak in apple trees for modeling dormancy]. *Semina: Ciências Agrárias,* 35(3): 1163-1176.

Asada, K. (2006). Production and scavenging of reactive oxygen species in chloroplast and their functions. *Plant Physiology,* 141(2): 391-396.

Aue, H. L., Lecomte, I. and Pétel, G. (1998). Activity and quantitative analysis of plasmalemma ATPase in buds and bud stands during winter vegetative bud's dormancy in peach-tree. *Acta Horticulturae,* 465: 531-537.

Aue, H. L., Lecomte, I. and Pétel, G. (2000). Changes in the parameters of the plasmalemma ATPase during peach vegetative bud dormancy. *Biologia Plantarum,* 43(1): 25-29.

Balandier, P., Gendraud, M., Rageau, R., Bonhomme, M., Richard, J.P. and Parisot, E. (1993). Bud break delay on single-node cuttings and bud capacity for nucleotide accumulation as parameters for endo- and paradormancy in peach trees in a tropical climate. *Scientia Horticulturae,* 55: 249-261.

Beauvieux, R., Wenden, B. and Dirlewanger, E. (2018). Bud dormancy in perennial fruit tree species: a pivotal role for oxidative cues. *Front. Plant Sci.* 9: 657.

Biasi, L. A., Carvalho, R. I. N. and Zanette, F. (2010). Dinâmica da dormência de gemas de videira e quivizeiro em região de baixa ocorrência de frio [Dormancy dymanics of grape and kiwifruit buds in a region of low chill occurrence]. *Revista Brasileira de Fruticultura,* 32(4): 1244-1249.

Biasi, L. A., Zanette, F. and Carvalho, R. I. N. (2012). Dormancy dymanics of grape and kiwifruit buds in a region of low chill occurrence. *Acta Horticulturae,* 932: 507-512.

Bonhomme, M., Rageau, R. and Gendraud, M. (2000). ATP, ADP and NTP contents in vegetative and floral peach buds during winter: are they useful for characterizing the type of dormancy? In *Dormancy in plants: From whole plant behaviour to cellular control*, edited by Viémont, J.-D.; Crabbé, J. 245-257. Cambridge: University Press.

Bonhomme, M., Rageau, R., Richard, J. P., Erez, A. and Gendraud, M. (1999a). Influence of three contrasted climatic conditions on endodormant vegetative and floral peach buds: analyses of their intrinsic growth capacity and their potential sink strength compared with adjacent tissues. *Scientia Horticulturae,* 80: 157-171.

Bonhomme, M., Rageau, R. and Gendraud, M. (1999b). Dynamics of ATP and NTP contents during the dormancy period in vegetative and floral buds of peach trees, influence of a cold deprivation. In: *International Symposium on Plant Dormancy,* 2, 1999, Angers, France. Book of Abstracts. Angers: Angers University, p.46.

Bonhomme, M., Rageau, R. and Gendraud, M. (1999c). Dynamics of carbohidrate contents of different tissues in and near vegetative and floral peach buds during dormancy period; influence of a cold deprivation. In: *International Symposium on Plant Dormancy,* 2, 1999, Angers, France. Book of Abstracts. Angers: Angers University, p.65.

Bowler, C., Montagu, M. V. and Inze, D. (1992). Superoxide dismutase and stress tolerance. *Annual Review of Plant Physiology and Plant Molecular Biology*, 43(1): 83-116.

Brasil. Ministério da Agricultura e Reforma Agrária. (2009). *Regras para Análise de Sementes* [*Rules for Seed Analysis*]. Secretaria Nacional de Defesa Agropecuária, Brasília.

Brown, D. S. and Kotob, F. A. (1957). Growth of flower buds of apricot, peach, and pear during the rest period. *Proceedings of the American Society for Horticultural Science,* 69: 158-164.

Buckhout, T. J. and Tubbe, A. (1996). Structure, mechanisms of catalysis, and regulation of sugar transporters in plants. In *Photoassimilate distribution in plants and crops: source-sink relationships*, edited by Zamski, E. and Schaffer, A. A. New York: Marcel Dekker, Inc, p.229-260.

Camelatto, D., Nachtigall, G. R., Arruda, J. J. P. and Herter, F. G. (2000). Efeito de flutuações de temperatura, horas de frio hibernal e reguladores de crescimento no abortamento floral de gemas de pereira [Effect of temperature fluctuations, chilling hours of winter and growth regulators on the floral abortion of pear tree buds]. *Revista Brasileira de Fruticultura,* 22(2): 111-117.

Carvalho, R. I. N. (2001). *Dinâmica da dormência e do conteúdo de carboidratos e proteínas em gemas vegetativas e ramos de um e dois anos de macieira com ou sem frio suplementar* [*Dynamics of dormancy and the content of carbohydrates and proteins in vegetative buds and branches of one and two years of apple tree with or without supplemental cold*]. 134 f. PhD dissertation in Plant Production. Universidade Federal do Paraná.

Carvalho, R. I. N. and Biasi, L. A. (2012). Índice para avaliação de intensidade de dormência de gemas de fruteiras de clima temperado [Index for bud dormancy evaluation of temperate climate fruit trees]. *Revista Brasileira de Fruticultura,* 34: 936-940.

Carvalho, R. I. N., Pereira, G. P., Prado, A. E., Biasi, L. A. and Zanette, F. (2012). Endodormancy of apple and pear buds in a region of low chill occurrence in Brazil. *Acta Horticulturae,* 932: 447-451.

Carvalho, R. I. N., Biasi, L. A., Zanette, F., Santos, J. M. and Pereira, G. P. (2010a). Estádios de brotação de gemas de fruteiras de clima temperado para o teste biológico de avaliação de dormência [Budburst stages of temperate zone fruit to the biological test of dormancy evaluation]. *Revista Acadêmica: Ciências Agrárias e Ambientais,* 8: 93-100.

Carvalho, R. I. N., Rendoke, J. C., Biasi, L. A. and Zanette, F. (2010b). Metabolic activity evaluation of temperate tree fruit buds by using the tetrazolium test. *Acta Horticulturae,* 872: 89-95.

Carvalho, R. I. N. and Silva, E. Q. (2010). Dormancy index of apple tree buds measured by the single node cutting biological test. *Acta Horticulturae,* 872: 101-105.

Carvalho, R. I. N. and Zanette, F. (2004). Dinâmica da dormência de gemas de macieira 'Imperial Gala' durante o outono e inverno em região de baixa ocorrência de frio [Dynamics of dormancy of 'Imperial Gala' apple buds during autumn and winter in a low cold region]. *Revista Brasileira de Fruticultura,* 26(1): 65-68.

Carvalho, R. I. N. and Zanette, F. (2005a). Morfologia de gemas dormentes com um ano e dois anos de macieiras cultivadas em regiões com insuficiência de frio hibernal [Morphology of dormant buds of apple trees with one year and two years old cultivated in regions with not enough winter cold]. *Revista Agropecuária Catarinense,* 18(2): 68-72.

Carvalho, R. I. N. and Zanette, F. (2005b). Variações do conteúdo de carboidratos em gemas e ramos de dois anos de macieira em região de baixa ocorrência de frio [Variations of the carbohydrate content in buds and branches of apple trees in a region of low cold occurrence]. *Revista Brasileira de Fruticultura,* 27(3): 352-355.

Chavarria, G., Herter, F. G. and Raseira, M. C. B. (2009). Effect of mild temperatures on bud breaking dormancy in low and medium chill peaches. *Ciência Rural,* 39(7): 2016-2021.

Dennis Junior, F. G. (2003). Problems in standardizing methods for evaluating the chilling requirements for the breaking of dormancy in buds of woody plants. *Hort Science,* 38(3): 347-350.

Dias, M. C. L. L. and Barros, A. S. R. (1995). *Avaliação da qualidade de sementes de milho* [*Evaluation of the quality of maize seeds*]. IAPAR, Londrina.

Erez, A., Faust, M. and Line, M. J. (1998). Changes in water status in peach buds on induction, development and release from dormancy. *Scientia Horticulturae,* 73(2-3): 111-123.

Erez, A. (2000). Bud dormancy: a suggestion for the control mechanism and its evolution. In: *Dormancy in plants: From whole plant behaviour to cellular control,* edited by Viémont, J.-D.; Crabbé, J. (ed.). Cambridge: University Press, p.23-33.

Farokhzad, A., Nobakht, S., Alahveran, A., Sarkhosh, A. and Mohseniazar, M. (2018). Biochemical changes in terminal buds of three different walnut (*Juglans regia* L.) genotypes during dormancy break. *Biochemical Systematics and Ecology,* 76: 52-57.

Guerriero, R., Gentili, M., Monteleone, P. and Viti, R. (2002). La valutazione dela dormienza nell'albicocco: tre metodi a confronto [The evaluation of dormancy with apricot: three methods compared]. *Rivista de Frutticoltura e di Ortofloricoltura,* 64(3): 73-78.

Hawerroth, F. J., Herter, F. G., Petri, J. L., Leite, G. B. and Pereira, J. F. M. (2010). Dormência em frutíferas de clima temperado [Dormancy in temperate fruit trees]. *Pelotas: Embrapa Clima Temperado*, 56 p. (Documentos, 310).

Hussain, S., Liu, G., Liu, D., Ahmed, M., Hussain, N. and Teng, T. (2015). Study on the expression of dehydrin genes and activities of antioxidative enzymes in floral buds of two sand pear (*Pyrus pyrifolia* Nakai) cultivars requiring different chilling hours for bud break. *Turkish Journal of Agriculture and Forestry,* 39: 930-939.

Khalil-Ur-Rehman, M., Wang, W., Xu, Y-S., Haider, M.S., Li, C-X. and Tao, J-M. (2017). Comparative study on reagents involved in grape bud break and their effects on different metabolites and related gene expression during winter. *Front. Plant Science*, 8: 1340.

Malagi, G., Robson, S. M., Citadin, I., Herter, F. G., Bonhomme, M., Regnard, J. L. and Legave, J.M. (2015). The comparison of dormancy dynamics in apple trees grown under temperate and mild winter climates imposes a renewal of classical approaches. *Trees,* 29: 1365-1380.

McPherson, H. G., Snelgar, W. P., Manson, P. J. and Snowball, A. M. (1997). Bud respiration and dormancy of kiwifruit (*Actinidia deliciosa*). *Annals of Botany,* 80(4): 411-418.

Meitha, K., Agudelo-Romero, P., Signorelli, S., Gibbs, D. J., Considine, J. A., Foyer, C. H. and Considine, M. J. (2018). Developmental control of hypoxia during bud burst in grapevine. *Plant Cell Environ.*, 41(5): 1154-1170.

Mittler, R. (2002). Oxidative stress, antioxidants and stress tolerance. *Trends in Plant Science*, 7(9): 405-410.

Mittler, R. (2017). ROS are good. *Trends in Plant Science*, 22(1): 11-19.

Mittler, R., Vanderauwera, S., Suzuki, N., Miller, G., Tognetti, V. B., Vandepoele, K., Gollery, M., Shulaev, V. and Van Breusegem, F. (2011). ROS signaling: the new wave? *Trends Plant Science,* 16: 300-309.

Or, E., Vilozny, I., Fennell, A., Eyal, Y. and Ogrodovitch, A. (2002). Dormancy in grape buds: isolation and characterization of catalase cDNA and analysis of its expression following chemical induction of bud dormancy release. *Plant Sci.*, 162: 121-130.

Pereira, G. P. (2017). *Fisiologia e bioquímica da dormência de gemas de caquizeiro 'Fuyu'* [*Physiology and biochemistry of dormancy of 'Fuyu' persimmon buds*]. 100 f. PhD dissertation in Plant Production. Universidade Federal do Paraná.

Pérez, F. J. and Lira, W. (2005). Possible role of catalase in post-dormancy bud break in grapevines. *Journal of Plant Physiology*, 162(3): 301-308.

Petri, J. L., Palladini, L. A., Schuck, E., Ducroquet, J. P., Matos, C.S. and Pola, A. C. (1996). *Dormência e indução da brotação de fruteiras de clima temperado* [*Dormancy and bud breaking of temperate fruit trees*]. EPAGRI, Florianópolis.

Pouget, R. (1963). *Recherches physiologiques sur le repos végétatif de la vigne (Vitis vinifera L.): la dormance des bourgeons et le mécanisme de sa disparition* [*Physiological research on the vegetative rest of the vine (Vitis vinifera L.): dormancy of the buds and the mechanism of its disappearance*]. 246 f. Thèse de Doctorat sciences naturelles, Université de Bordeaux, Bordeaux.

Putti, G. L., Petri, J. L. and Mendez, M. E. (2003). Temperaturas efetivas para a dormência da macieira (*Malus domestica* Borkh.) [Effective temperatures for apple dormancy (*Malus domestica* Borkh.)]. *Revista Brasileira de Fruticultura* 25(2): 210-212..

Raven, P. H., Evert, R. F. and Eichhorn, S. E. (2001). *Biologia vegetal* [*Plant Biology*]. Guanabara Koogan, Rio de Janeiro.

Sudawan, B., Chang, C-S., Chao, H-F., Ku, M. S. B. and Yen, Y-F. (2016). Hydrogen cyanamide breaks grapevine bud dormancy in the summer through transient activation of gene expression and accumulation of reactive oxygen and nitrogen species. *BMC Plant Biology,* 16: 202.

Tabuenca, M. C. (1964). Necesidades de frío invernal de variedades de albaricoquero, melocotonero y peral [Chilling requirements of apricot, peach and pear varieties]. *Anales de la Estación Experimental de Aula Dei,* 7(3/4): 113-132.

Tabuenca, M. C. (1967). Necesidades de frío invernal de variedades de ciruelo [Chilling requeriments of plum varieties]. *Anales de la Estación Experimental de Aula Dei,* 8: 383-391.

Tabuenca, M. C. (1968a). Necesidades de frío invernal de variedades de peral [Chilling requeriments of pear varieties]. *Anales de la Estación Experimental de Aula Dei,* 9(1): 1-9.

Tabuenca, M. C. (1968b). Necesidades de frío invernal de variedades de albaricoquero [Chilling requeriments of apricot varieties]. *Anales de la Estación Experimental de Aula Dei,* 9(1): 10-24.

Tabuenca, M. C. (1969). Necesidades de frío invernal de variedades de melocotonero [Chilling requeriments of peach varieties]. *Anales de la Estación Experimental de Aula Dei,* 10(4): 946-956.

Tabuenca, M. C. (1972a). Necesidades de frío invernal de variedades de almendro [Chilling requeriments of almond varieties].. *Anales de la Estación Experimental de Aula Dei,* 11(3/4): 325-329.

Tabuenca, M. C. (1972b). Necesidades de frío invernal de variedades de melocotonero. II [Chilling requeriments of peach varieties. II]. *Anales de la Estación Experimental de Aula Dei,* 11(3/4): 330-334.

Tabuenca, M. C. (1976a). Necesidades de frío invernal de variedades de nectarina [Chilling requeriments of nectarine varieties]. *Anales de la Estación Experimental de Aula Dei,* 11(3/4): 256-260.

Tabuenca, M. C. (1976b). Inflencia del patrón e de la temperatura en el periodo de reposo invernal de variedades de albaricoquero [Influence of the rootstock and the temperature on the rest period of apricot varieties]. *Anales de la Estación Experimental de Aula Dei,* 11(3/4): 325-338.

Tabuenca, M. C. (1976c). Reposo invernal de ciruelos e manzanos empleados como patrones [Rest period of plum and apple rootstocks].. *Anales de la Estación Experimental de Aula Dei,* 11(3/4): 339-344.

Tamura, F., Tanabe, K., Itai, A. and Tanaka, H. (1998). Protein changes in the flower buds of Japanese pear during breaking of dormancy by chiling or high-temperature treatment. *Journal of the American Society for Horticultural Science,* 123(4): 532-536.

Takemura, Y., Kuroki, K., Jiang, M., Matsumoto, K. and Tamura, F. (2015). Identification of the expressed protein and the impact of change in ascorbate peroxidase activity related to endodormancy breaking in *Pyrus pyrifolia*. *Plant Physiology and Biochemistry,* 86: 121-129.

Thorpe, M. R. and Minchin, P. E. H. (1996). Mechanisms of long- and short-distance transport from sources to sinks. In *Photoassimilate distribution in plants and crops: source-sink relationships,* edited by Zamski, E. and Schaffer, A. A.. New York: Marcel Dekker, Inc,. p.261-282.

Vergara, R., Parada, F., Sebastián, R. and Pérez, F. J. (2012). Hypoxia induces H2O2 production and activates antioxidant defence system in grapevine buds through mediation of H2O2 and ethylene. *Journal of Experimental Botany*, 63(11): 4123-4131.

Viti, R., Bartolini, S., and Zanol, G. C. (2012). Biological changes and active oxygen-scavenging enzymes activities in apricot (*Prunus armeniaca* L.) flower buds during dormancy transitions. *Acta Horticulturae,* 940: 331-339.

Wagner Junior, A., Bruckner, C. H., Pimentel, L. D. and Raseira, M. C. B. (2006). Evaluation of chilling requirement in peach through grafted twigs. *Acta Horticulturae,* 713: 243-246.

Wang, S. Y., Faust, M. and Jiao, H. J. (1991). Changes in superoxide dismutase activity during thidiazuron-induced lateral budbreak of apple. *Hort Science,* 26(9): 1202-1204.

Yaacoubi, A. E., Malagi, G., Oukabli, A., Citadin, I., Hafidi, I., Bonhomme, M. and Legave, J. M. (2016). Differentiated dynamics of bud dormancy and growth in temperate fruit trees relating to bud phenology adaptation, the case of apple and almond trees. *Int J Biometeorol,* 60: 1695-1710.

In: Plant Dormancy
Editor: Renato V. Botelho

ISBN: 978-1-53615-380-4

Chapter 4

MANAGEMENT OF DORMANCY RELEASE ON DECIDUOUS FRUIT TREES UNDER MILD WINTER CONDITIONS

***José Luiz Petri*[1,*], *Fernando Hawerroth*[2] and *Andre Amarildo Sezerino*[1]**

[1]Epagri – Agricultural Research and Rural Extension Company, Caçador, Santa Catarina State, Brazil

[2]Embrapa - Brazilian company of agricultural research, Vacaria, Rio Grande do Sul State, Brazil

ABSTRACT

The best and most efficient way to avoid problems related to dormancy on deciduous fruit trees is the use of cultivars adapted to the amount of winter cold in the growing region. In addition to the use of cultivars with low chilling requirement, there are some technical interventions that can minimize the effects of lack of chilling on low sprouing and flowering, such as artificial chilling, ring incision, arching of branches, growth

* Corresponding Author's E-mail: petri@epagri.sc.gov.br.

control, post-harvest pruning and defoliation. After the nursery trees are removed from the soil, they can be stored in a cold room for a period of 30 to 45 days at 2 to 6 °C, allowing a uniform sprouting and greater branches growth. The ringing can be used, in the first and second year, for the plant in formation of the upper layers when a suitable sprouting for the plant formation did not occur. In the climatic conditions of southern Brazil, the period of vegetative development of temperate fruit trees is superior to that observed in typical regions of temperate climate, which associated with high temperatures and high rainfall rates during the cycle can result in excessive branch growth. Under such conditions, the use of management techniques that restrict vegetative development becomes essential. Among the measures to control the vegetative growth of the branches of the year, the use of calcium prohexadione is efficient in apple, pear and plum tree crops. In mild climatic conditions the temperate fruit trees resume the vegetative growth after harvesting, under these conditions pruning when the plants still have leaves reduces the plant vigor and photoassimilates are reallocated to the reproductive organs that will remain in the plant. Arching provides a weakening of the branches, favoring sprouting. Branches ringed to interrupt the normal circulation of sap also favor sprouting. In regions of subtropical climate or in years that the temperatures are not too low in the autumn-winter, the natural fall of leaves does not occur. Under these conditions it is necessary to make chemical defoliation, which should preferably be carried out with chemicals that promote leaf senescence.

Keywords: mild winter, plant regulators, sprouting, dormancy release

1. Introduction

The best and most efficient way to avoid problems related to dormancy on deciduous fruit trees is the use of cultivars adapted to the amount of winter chilling in the growing region. In regions of mild winter, preference should be given to low-requirement chilling cultivars. As these cultivars generally flower earlier, care should be taken to grow them in places where intense spring frost is not frequent. For most species, plant cultivars with chilling requirements below 400h ($\leq$ 7.2°C) were developed. Among these species stand out blackberry, raspberry, peach, plum, apple, pear, quince and kiwi trees. In addition to the use of cultivars with low chilling requirement, there are some technical interventions that can minimize the effects of lack of chilling on low sprouging and flowering, such as artificial chilling, ring incision, arching of branches, growth control, post-harvest pruning and defoliation.

2. ARTIFICIAL COLD TO INDUCE SPROUTING IN NURSERY TREES

After the nursery trees are removed from the soil, they can be stored in a cold room for a period of 30 to 45 days at 2 to 6°C, allowing a uniform budding and greater branches growth (Tables 1 and 2, Figure 1). It is observed that temperatures between 2 and 6°C show small differences in the percentage of sprouting of the buds and in the growth of the branches, depending on the period of storage. Therefore, it is recommended to store at temperatures of 4 to 6°C. When kept only for 30 days, it is recommended to complement with the chemical treatment with sprouting inducers to increase the sprouting percentage (Tables 3). The storage of nursery trees for periods superior to 30 days allows an increase in branches growth, allowing a better plant formation (Figure 1). This greater growth provides increase of the plant vigor, with greater leaf area, and a greater root development.

Table 1. Effect of temperature and storage period in cold chamber on dormancy of apple nursery trees, cultivar Gala - Caçador, Santa Catarina State, Brazil

Days	Percentage of sprouted buds					
	Year 1		Year 2		Average	
	2°C	6°C	2°C	6°C	2°C	6°C
0	24.0	29.5	17.9	21.3	20.9	25.4
15	28.5	21.0	31.1	32.3	29.8	26.6
30	60.5	48.0	37.5	36.2	49.0	42.1
45	81.5	68.5	60.0	57.6	70.7	63.0
60	70.0	79.5	86.7	69.6	78.3	74.5

During storage period, it is important to maintain relative humidity above 80% and periodically, if necessary, wet the roots. The nursery trees should not be placed together with fruits, because the ethylene produced and released impairs a proper sprouting of buds.

Table 2. Effect of temperature and storage period in cold room on the dormancy of apple nursery trees, cultivar Fuji - Caçador, Santa Catarina State, Brazil

Days	Percentage of sprouted buds					
	Year 1		Year 2		Average	
	2°C	6°C	2°C	6°C	2°C	6°C
0	40.0	43.0	16.0	15.0	28.0	29.0
15	62.0	46.0	27.0	23.0	44.5	34.5
30	83.0	83.0	33.0	29.0	58.0	56.0
45	97.0	88.0	55.0	46.0	76.0	67.0
60	91.0	89.0	78.0	56.0	84.5	72.5

Table 3. Effect of storage period in cold room with and without chemical treatment for dormancy breaking of apple nursery trees, cultivar Fuji - Caçador, Santa Catarina State, Brazil

Days in Cold chamber	Percentage of sprouted buds			
	Year 1		Year 2	
	Without chemical treatment	With chemical treatment	Without chemical treatment	With chemical treatment
0	19.0	63.0	12.0	19.0
15	32.0	76.0	16.0	34.0
30	77.0	90.0	21.0	41.0
45	87.0	98.0	40.0	61.0
60	85.0	96.0	58.0	77.0

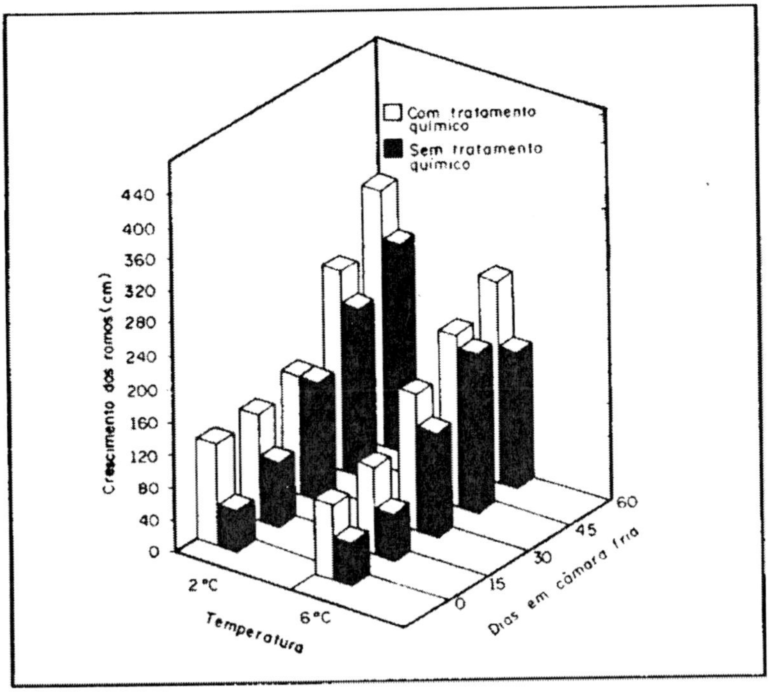

Figure 1. Effect of storage period and temperature on budding.

Figure 2. Apple nursery trees that did not receive enough chilling (left) or that remained 45 days in the cold room (right) with better sprouting and higher growth.

3. RINGING

The ringing can be used, in the first and second year, for the plant in formation of the upper layers when a suitable sprouting for the plant formation did not occur. It consists of the interruption of sap circulation, which transports growth inhibitors, thus stimulating budding. This procedure is done by removing the bark one centimeter above the buds, soon after beginning of sprouting and also performing a cut of the bark in vertical direction, just below the bud, with the purpose of stimulating the sprouting of branches with greater angle of insertion (Figure 3).

4. CONTROL OF VEGETATIVE GROWTH

The growth cycle of temperate fruit trees under mild winter conditions or subtropical climate is altered, prolonging the growth phase and delaying dormancy (Figure 4).

NOTE: A - Ring incision in dormant bud; B - Sprouting of the bud that received the girdling. Photos: Gabriel Berenhauser Leite and Rafael Pio.

Figure 3. Ringing in apple tree seedlings.

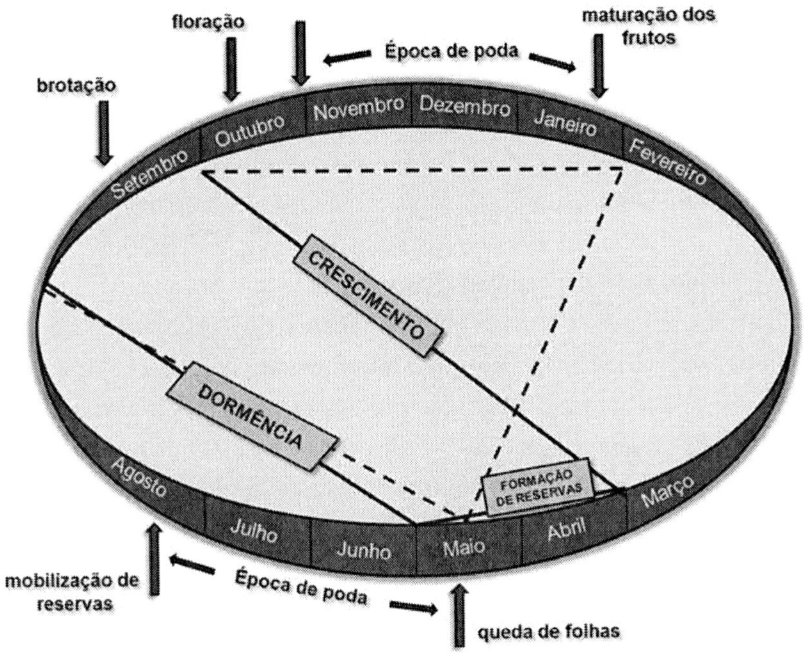

Figure 4. Vegetative cycle of temperate fruit trees in different climatic conditions. Dashed line - characteristically temperate climate conditions (cold winter); Continuous line - subtropical to temperate (mild winter) conditions. Caçador, Santa Catarina State, Brazil.

Excessive vegetative development in fruit species can compete with fruit growth (Basak e Rademacher 2000), affecting negatively the fruiting by the decrease of the number of cells per fruit, limiting the fruit growth (Yamaguchi et al. 2002).

In addition, vigorous shoot growth reduces the distribution of light within the canopy (Privé et al. 2004), adversely affecting fruit quality and disease control.

In the climatic conditions of southern Brazil, the period of vegetative development of temperate fruit trees is superior to that observed in typical regions of temperate climate, which associated with high temperatures and high rainfall rates during the cycle can result in excessive branch growth. Under such conditions, the use of management techniques that restrict vegetative development becomes essential.

Preferably, control of vegetative development should be performed by natural means, such as the use of less vigorous rootstocks, the use of formation and fruiting pruning. However, if these solutions can not be implemented or are not sufficient to restrict canopy vigor, the reduction of vegetative development can be achieved by the use of plant regulators. Control of vegetative growth by chemical means can help reduce excessive growth by limiting plant size or restricting growth at a given time, allowing a better balance between vegetative development and fruiting (Miller 2002).

In fruit growing, the main plant regulators that make can restrict the vegetative development are the chlormequat chloride, daminozide, ethephon, paclobutrazol, calcium prohexadione and trinexapac-ethyl. According to Rademacher (2000), the plant regulators that act in the diminution of the vegetative development can be divided into inhibitors of the synthesis of gibberellins and compounds that induce the formation of ethylene.

The application of ethephon can reduce the growth of branches. Ethephon, when applied to plant tissues, is degraded into ethylene. According to Wertheim and Webster (2005), the formation of ethylene in tissues possibly induces a decrease in the availability of auxins in the branches terminal regions, triggering the reduction of their growth. Considering the effects of ethephon on the abscission and fruit maturation due to increase of ethylene synthesis, the use of ethephon for control of vegetative development has not been used. To improve efficiency of ethephon (50 to 100mg L^{-1}) it is mixed with naphthalene acetic acid (10 to 20mg L^{-1}).

Paclobutrazol has been widely reported as an inhibitor of shoot growth in pear trees (Dheim And Browning 1988; Sansavini et al. 1988, Browning et al. 1992) and in apple trees (Camilo 2006). according to camilo (2006), in addition to its effect on suppression of plant growth, paclobutrazol may have an adverse effect, reducing fruit size in the year of application. According to this author, the effect of paclobutrazol on suppression of terminal and lateral growth of the branches and reduction of fruit size may also be prolonged for the year following application. Paclobutrazol, due to its residual effect on both soil and plant, is rarely used in temperate fruit trees. Paclobutrazol presents as great disadvantage its great persistence in plants and high toxicity (Greene 1986; Steffens e Wang 1986).

The calcium prohexadione has good results in apple trees, but when used in high concentrations in pear trees, can cause negative effects on flowering (Deckers and Schoofs 2004). The most significant effect of the application of high concentrations of calcium prohexadione is the reduction of floral return in the following year (Sugar et al. 2002; Meintjes et al. 2005; Asin et al. 2005).

Calcium prohexadione, in the form of free acid, exhibits structural similarities with 2-oxoglutaric acid and ascorbic acid and, consequently, inhibits distinct dioxigenases that require these compounds as co-substrates (Costa et al. 2006; Rademacher 2009).

A B

Figure 5. Effect of calcium prohexadione on the vegetative growth of the apple tree. A) With application; B) No Application.

According to Rademacher (2009), the dioxigenases being blocked catalyze reactions of biosynthesis of gibberellins of lesser degree of activity, decrease the synthesis of ethylene, and alter the biosynthesis of flavonoids. Ethylene is generated from 1-carboxy-1-amino-cyclopropane (ACC) in a reaction catalyzed by ACC oxidase, an enzyme that requires ascorbic acid as co-substrate.

Among the measures to control the vegetative growth of the branches of the year, the use of calcium prohexadione is efficient in apple, pear and plum tree crops (Figure 5 and Table 4).

Copper chloride oxide may also be used during the vegetative phase at low dosages or post-harvest, which provides a better lignification of growing branches.

Table 4. Alternatives of application time and concentration of Viviful® in the apple tree crop

Time of application	Viviful® Concentration	Observations
1. Full bloom and petal fall	400 to 800 g ha^{-1}	Control of growth and increase of effective fruiting. Reapply after 30 days
2. Branches 5 to 15 cm	1.200 g ha^{-1}	Resuming the growth of the branches, reapply.
3. Branches with 5 cm 30 days after first application	600 g ha^{-1} 600 g ha^{-1}	Application at 30 days after, only if the growth of the branches has reassumed.
4. Branches 5 cm 30 days after first application 60 days after first application	400 g ha^{-1} 400 g ha^{-1} 400 g ha^{-1}	Reapply at 30 days after the first application and at 60 days only if the growth of the branches is reassumed.

5. Pruning after Harvest

In mild climatic conditions the temperate fruit trees resume the vegetative growth after harvesting, forming weak buds in the branches, and these branches do not always come to lignify and can suffer damage when using sprouting inducers.

Under these conditions pruning when the plants still have leaves reduces the plant vigor and photoassimilates are reallocated to the reproductive organs that will remain in the plant.

Post-harvest pruning is performed whenever there is excessive branches growth, shortly after harvest, and can be performed in different species of temperate fruit trees (Figure 6).

6. Arching Branches

Arching provides a weakening of the branches, favoring sprouting. Branches ringed to interrupt the normal circulation of sap also favor sprouting.

Figure 6. Pruning after peach harvest.

A B

C

Figure 7. Arched branches with the consequent formation of spurs.

The arching branches aims to give a direction to the branches and control the vigor of the same, allowing for better light to enter inside the plant allowing a better formation of flower buds and control of apical dominance.

When apical dominance occurs, buds of terminal growth inhibits sprouting of axillary buds, since the auxins are produced at the apex and move by gravity, generating an imbalance in relation to cytokinins, responsible for sprouting of axillary buds (Hoffmman et al. 2004).

Figure 8. Induced senescence in apple plants.

The arching of the branches is the most natural and most used process and consists in directing the branches to angulation from 45° to 120° and can be done with a stick, ribbons or wood. This practice can be used when part of the branch of the year is 10cm to 80cm. (Figure 7).

7. DEFOLIATION

In regions of subtropical climate or in years that the temperatures are not too low in the autumn-winter, the natural fall of leaves does not occur. Under these conditions it is necessary to make chemical defoliation, which should preferably be carried out with chemicals that promote leaf senescence (Figures 8 and 9). It can be made with 8% to 10% urea, 1% copper sulfate, 0.2% copper oxychloride, 0.1% Ethrel 240 and 0.1% Abscisic acid; two months before the normal start of sprouting (Table 5).

As for the products based on copper, they do not allow the leaves senescence, but their burning with consequent fall. This practice should also be used in the nursery.

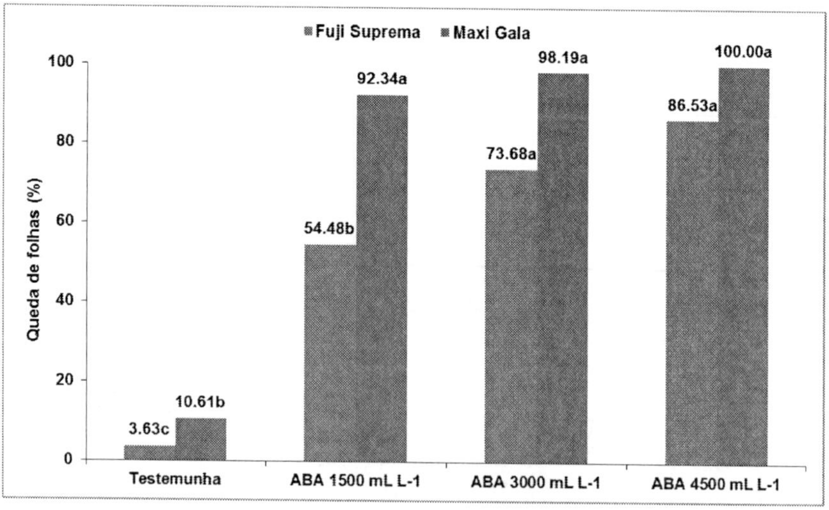

Figure 9. Percentage of leaf fall ten days after the application of Abscisic Acid (AB) in cultivars Fuji Suprema and Maxi Gala. Caçador, Santa Catarina State, Brazil.

Table 5. Effect of chemical products on percentage of defoliation of apple tree, cv. Fuji Supreme in branches up to 30 cm in length. Caçador, Santa Catarina State, Brazil, 2003

TREATMENTS	% of leaf fall				
	28/06	08/07	15/07	22/07	06/08
KNO_3 – 10%	22.9ns	72.8b	73.5b	75.7b	85.6c
UREA – 10%	27.3	98.1a	98.4a	98.4a	99.7a
FITOFOS – K – 3%	28.4	71.1b	71.7b	73.1b	93.1b
COPPER – Cu – 5%	22.6	56.8c	58.7c	69.4b	93.5b
CONTROL	28.3	49.2d	50.1d	57.1c	87.4c

Fuji Long branches 50cm

TREATMENTS	% of leaf fall				
	28/06	08/07	15/07	22/07	06/08
KNO_3 – 10%	20.5	57.8	58.4	59.9	68.6
UREA – 10%	20.1	82.0	82.7	83.1	86.5
FITOFOS – K – 3%	20.6	72.5	73.5	73.7	77.3
COPPER – Cu – 5%	17.9	43.6	43.8	50.9	93.1
CONTROL	17.9	42.8	44.2	48.8	93.5

REFERENCES

Asín, L.; Dalmau, R.; Bonany, J.; Pages, J. M.; Vilardell, P. (2005). Effect of Proexadione-Ca on growth regulation, yield, fruit set and return bloom in 'Blanquilla' and 'Conference', the two main cultivars in Spain. *Acta Horticulturae,* 671: 525 - 532.

Basak, A.; Rademacher, W. (2000). Growth regulation of pome and stone fruit trees by use of proexadione-Ca. *Acta Horticulturae,* 514: 41-51.

Browning, G.; Kuden, A.; Blake, P. 1992. Site of (2RS, 3RS)-paclobutrazol promotion of axillary flower initiation in pear cv. Doyenne du Comice. *Journal of Horticultural Science,* 67 (1): 121-128.

Camilo, A. P.; Pereira, A. J. (2006). Thinning fruits (raleio de frutos). In: EPAGRI. *The apple tree cultivation* (*A cultura da macieira*), Florianópolis, p. 419 - 461.

Costa, G.; Dal Cin, V.; Ramina, A. (2006). Physiological, molecular and practical aspects of fruit abscission. *Acta Horticulturae*, 727: 301-310.

Deckers, T., H. Schoofs And E. Smolders. (2005). "Growth Regulation - Natural or Chemical Growth Regulation in Pear". *Acta Horticulturae,* 671: p.503.

Dheim, M. A.; Browning, G. (1988). The mechanism of the effect of (2RS, 3RS)-paclobutrazol on flower initiation of cvs Doyenne du Comice and Conference. *Journal of Horticultural Science*, 63 (3): 393-405.

Greene, D. W. (1986). Effect of paclobutrzol and analogs on growth, yield, fruit quality, and storage potential of 'Delicious' apples. *Journal of American Society for Horticultural Science,* 111: 328-332.

Hoffmann, A.; Petri, L. L.; Leite, G. B.; Zancan, C.; Camelatto, D. (2004). Cultural Practices. In: *Apple: Production*. Nachtigall, Gilmar Ribeiro, Embrapa Uva e Vinho; Brasilia: Embrapa Informação Tecnologica, 2004.

Meintjes, J. J.; Stassen, P.; Theron, K. J. (2005). The effects of different rates of Prohexadione-calcium and girdling on shoot growth and fruit quality when applied to different pear cultivars. *Acta Horticulturae,* 671: 539-546.

Miller, S. S. Prohexadione-calcium controls vegetative shoot growth in apple. (2002). *Journal of Tree Fruit Production*, 31 (1): 11-28.

Privé, J. P. Fava, E., Cline, J. E., Byl, M. (2004). Preliminary results on the efficacy of apple trees fruit with the growth retardant Prohexadione–Calcium (Apogee) in the Eastern Canada. *Acta Horticulturae,* 636: 137-144.

Rademacher, W. (2009). Prohexadione-Ca – A new plant bioregulator for use in apple production. In: Encontro sobre Fruticultura de Clima Temperado, 11, 2009, Fraiburgo, SC, *Anais...,* Caçador: Epagri, v.1 (Palestras), p. 1-10.

Sansavini, S.; Cristoferi, G.; Montalti, P. (1988). Effects of Paclobutrazol on growth, fruiting, carbohydrate metabolism in pear trees. *Advances in Horticultural Science,* 2: 52-57.

Sugar, D.; Elfing, D. C.; Mielke, E. A. (2002). Effects of prohexadione-calcium (Apogee™) on blossoming, production and fruit quality in pear. *Acta Horticulturae*, 596: 757-760.

Steffens, G. L.; Wang, S. Y. (1986). Biochemical and physiological alterations in apple trees caused by a gibberellin biosynthesis inhibitor, paclobutrazol. *Acta Horticulturae,* 179: 433-442.

Weirtheim, S. J.; Webster, A. D. (2005). Manipulation of growth and development by plant bioregulators. In: *Fundamentals of temperate zone tree fruit production*. Leiden: Backhuys, p.267-294.

Yamaguchi, M.; Haji, T.; Miyake, M.; Yaegaki, H. (2002). Varietal difference in cell division and enlargement periods during peach (*Prunus persica* Batsch) fruit development. *Journal of Japanese Society for Horticultural Science,* 71: 155-163.

In: Plant Dormancy
Editor: Renato V. Botelho

ISBN: 978-1-53615-380-4

Chapter 5

PHYSIOLOGY AND MANAGEMENT OF VINES DORMANCY IN TROPICAL AND SUBTROPICAL CONDITIONS

Alessandro Jefferson Sato[1,]*
Sergio Ruffo Roberto,[2]
and Renato Vasconcelos Botelho[3]
[1] Department of Agronomic Science, Federal University of Paraná, Paraná State, Brazil
[2] Department of Agronomy, Center of Agrarian Sciences, State University of Londrina, Londrina, Paraná State, Brazil
[3]State University of the MidWestern of Paraná (Unicentro), Guarapuava, Paraná State, Brazil

ABSTRACT

Hibernal dormancy of vines are characterized by a plant state of physiological inactivity, which allows its survival under adverse

* Corresponding Author's E-mail: asato@ufpr.br.

conditions. In this period, the plant does not show vegetative growth, but its metabolic activities continues, although reduced. For these vines to resume the growth naturally, its exposure to a certain period of low temperatures is required, called chilling units. The dormancy presents progressive development that begins during the autumn, increasing its intensity until reaching the deep dormancy or endodormancy, conferring the capacity of survival in temperatures unsuitable to vegetal development. Thus, one of the major limitations for grapevine cultivation in tropical and subtropical regions, where the amount of hibernal chilling is not enough, is the overcoming of your period of dormancy. In these regions, the application of some substances is a strategy that can be used to induce bud burst, flowering and ripening of temperate climate fruit species like vine. Among several substances that have an effect on artificial induction of bud burst, we highlight the hydrogen cyanamide, successfully used in several vine growing regions, as well as other promising products such as mineral and vegetable oils and garlic extract. This chapter approaches the main aspects related to grapevine dormancy, as well as the strategies for induction of a new cycle in tropical and subtropical regions through application of bud burst inducers.

Keywords: dormancy, budbreak, grapevine, off-season crop, plant growth regulators

1. Introduction

One of the vines characteristics, as well as other species of temperate climate, is the hibernal dormancy, a complex phenomenon governed by various biochemical and enzymatic processes, and dependent on the plant reserve substances, their nutritional and hormonal status. The dormancy can be considered a physiological inactivity state of the vine that allows its survival in adverse conditions, as the low temperatures during the winter. In this period, the plant does not show growth, but its metabolic activities continue, although reduced (Figure 1.1). In order for these vines to resume growth naturally, their exposure to a certain period of low temperatures is required, variable for each cultivar (Petri et al. 2002; Pires; Pommer 2003; El-Yazal et al. 2014).

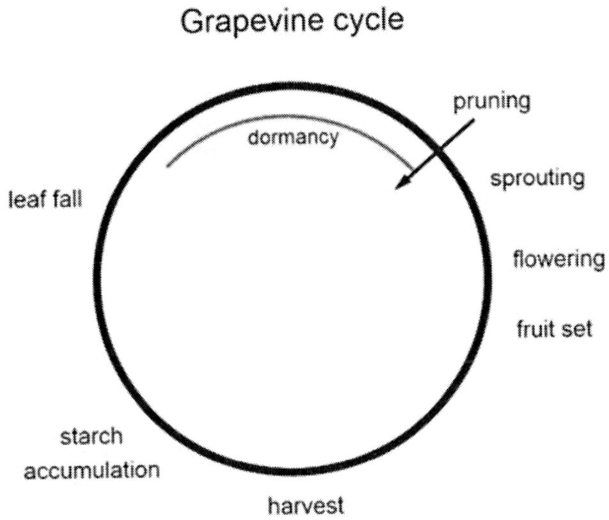

Figure 1. Grapevine cycle throughout the year.

During the occurrence of this phenomenon, large changes in hormonal balance of the plant are verified. One of the plant regulators that can influence much of this process is abscisic acid (ABA), which acts as primary regulator in the initiation of seed and bud dormancy and in plant responses to stress, particularly hydric stress. In addition, ABA influences various aspects of plant development, acting as an antagonist to the action of growth promoters, such as gibberellins, auxins and cytokinins, and inhibiting certain types of RNA (ribonucleic acid), which prevent formation of proteins necessary for growth. However, positive correlations between ABA content, number of chilling hours and dormancy of buds of different species could led to erroneously consider the direct involvement of this compound in control of buds dormancy (Saure 1985; Powell 1987; Petri et al. 2006; El-Yazal et al. 2014; Pinto et al. 2007).

Although dormancy is extensively studied, the knowledge of the physiological mechanisms involved in this process is still limited, due to the large number of environmental and plant factors involved, as well as methodological difficulties in studies of this nature (Olsen 2006; Hawerroth et al. 2010; Campoy et al. 2011).

It is known that dormancy is a mechanism that is acquired as a process of progressive development that begins during autumn, increasing its intensity until reaching the so-called deep dormancy or endodormancy, conferring the capacity of survival at temperatures unsuitable for plant development (Lang et al. 1987; Powell 1987; Ruiz et al. 2007).

This process is characterized by three successive periods: paradormancy, endodormancy and ecodormancy. Paradormancy, also called correlative inhibition, is the result of the influence of another plant organ on the buds, which causes inactivation of the floral or vegetative meristem. An example of this is the apical dominance, caused by basipetal flow of auxin, which inhibits bud burst of lateral buds. In the case of vines, this phase occurs soon after the differentiation of the grape bud's meristems into reproductive or vegetative ones. In this period, many buds, especially the basal ones, could burst, but remain at rest, mainly due to the dominance exerted by apical buds (Lang et al. 1987; Pinto et al. 2007; Lavee; May 1997).

During paradormancy, the buds lose chlorophyll and become light to dark brown, due to the accumulation of lignin and other phenolic compounds in the cell walls. Subsequently, endodormancy is initiated as a way of survival in environmental conditions unfavorable to growth, usually reduction of photoperiod and fall in temperatures. When the buds are at this stage, exposure to optimal developmental conditions is not sufficient to induce their sprouting. The buds must be exposed before to environmental conditions that stimulate the endodormancy state to recover their sprouting capacity (Horvath et al. 2003; Perez; Lira 2005; Haverroth et al. 2010).

Another factor related to bud dormancy is ecodormancy that keeps buds in a quiescent state, even if it is physiologically apt to develop, due to environmental factors limiting the development, that is, the bud burst will occur only after the establishment of favorable climatic conditions, especially temperature (Horvath et al. 2003; Haverroth et al. 2010).

In this way, the environment temperature is a determining factor in plants of temperate climate dormancy, either for its induction or for its overcoming. If the plants are dormant, the continuous action of low temperatures for a certain period allows the release of dormancy. This amount of chilling hours is variable among species, cultivars and even buds within the same plant (Petri et al. 2006; Campoy et al. 2011).

The chilling is the main responsible for the accumulation of oxidizing compounds in the buds, especially the hydrogen peroxide (H_2O_2), which is directly related to dormancy overcoming in vines because it is mediated by highly reactive O_2 free radicals. H_2O_2 is synthesized when the plant is exposed to a biotic or abiotic stress situation (Prasad et al. 1994; Gárcia et al. 2007).

The increase in H_2O_2 levels initiates a process of signal transfer, resulting in the end of endordomancy phase, initiating bud burst. Therefore, it can be affirmed that the endordomancy ends by the accumulation of H_2O_2 in the tissues. This accumulation or oxidative stress is caused by the inhibition of catalase activity or by the action of peroxidases that oxidize NADH (nicotinamide adenine dinucleotide). Both phenomena are stimulated by exposure to cold or action of chemicals, such as hydrogen cyanamide (Gemma 1995).

Catalase is an enzyme present in aerobic cells that decomposes H_2O_2 into O_2 and H_2O, which has the physiological role of preventing accumulation of peroxide and subsequent, cellular damage. The activity of this enzyme is reduced when the vine is in endodormancy. All of these metabolic alterations would result in an increase in the levels of the intracellular AMP / ATP ratio that would induce the expression of protein-kinases, which would form part of the signal transfer system that would lead to the end of the endodormancy. Apparently, genes related to biosynthesis of plant regulators, such as gibberellins and cytokinins, are activated during this process, since bud burst begins with the cell division process, breakage of starch and growth, generally, regulated by these hormones (Prasad et al. 1994; Gemma, 1995).

The productivity of a vine orchard depends, in addition to other factors, on obtaining a high index of bud burst. To achieve this, the buds are usually treated with specific products. The factors related to plant and environment also have a great influence on sprouting.

Normally, well-replenished plants sprout better than those weakened and suffer less from the climatic oscillations, so frequent in the winter of regions of subtropical climate. High temperatures after treatments advance the onset of sprouting and decrease delay.

In such climate conditions, instead of "breaking dormancy," it is better to use the term "forcing or inducing bud burst" because, in practice, to obtain the early crop, buds are also stimulated in the summer, when the plant is in full vegetative activity.

2. Induction of Bud Burst in Grapevines

One of the major limitations of grapevine production in tropical and subtropical regions is to overcome its dormancy period, because the chilling accumulation is not enough (Kishino et al. 2007). In these cases, the incomplete overcoming of the rest determines the delay and low percentage of bud burst, and lack of uniformity in the flowering of plants, and as a result, there is a reduction in the production and quality of the fruits (Petri et al. 1996; Or et al. 2000).

However, favorable environmental conditions, such as high temperatures, are required for bud burst after the dormancy period has elapsed. The beginning of the annual growth cycle is closely related to the environment temperature since it acts on the metabolic speed processes that promote bud burst (Mullins et al. 1992).

The use of promoter agents is a strategy that can be used to induce bud burst, flowering and ripening of fruits of temperate climate fruit species such as grapevines. The first reference of the action of products on bud burst induction dates back to 1918, when it was found that apple trees sprayed with linseed oil mixed in water during dormancy to combat aphids, flourished first and for a shorter period (George et al. 2002; Petri et al. 2006).

Even in places where dormancy is naturally overcome, the use of bud burst inducers can be beneficial, since it is possible to predict the time of harvest of the fruits, to obtain greater bud burst uniformity and to increase the number of bursted buds. Many chemical agents have efficiency in sprouting induction, but few of them are commercially accepted substances. Among the various substances that have an effect on bud sprout induction, we can highlight: mineral and vegetable oil, calcium cyanamide, potassium nitrate, hydrogen cyanamide, dinitro-ortho-cresol (DNOC), dinitro-ortho-butyl-phenol (DNOPB), dinitrobutylphenol (DNBP), thiourea, sodium pentachlorophenolate, thidiazuron (TDZ), garlic extracts and gibberellic acid. It should be noted that dinitro salts were banned from the market in the 1980s because of their high toxicity (Erez 2000; Haverroth et al. 2010).

2.1. Hydrogen Cyanamide and Calcium Cyanamide

In Brazil, hydrogen cyanamide has been used as the main strategy to induce bud burst of grapevines grown in tropical and subtropical regions. However, before its discovery, calcium cyanamide was initially used with great success in the induction of bud burst, and it is still used by some growers of 'Niagara Rosada' grapes.

Calcium cyanamide is an organic nitrogen fertilizer containing 63% calcium cyanamide ($CaCN_2$), 17% $Ca(OH)_2$ and 12% free carbon. The most used powder product has 22% nitrogen, 98% of which is soluble in water (Malavolta 1981).

The first result on the positive effect of the calcium cyanamide solution on bud burst in 'Italia' table grape in Brazil was obtained in 1969 by the agronomist Masasuke Mashima from the Agricultural Cooperative COTIA in Londrina, Paraná State. It was found that the effect of calcium cyanamide was limited only to treated buds (Kuroi; Shiraishi; Mano 1963; Kuroi 1974) and, therefore, its application should be directed exclusively to the vine buds. Later, the researcher Antonio Yoshio Kishino, from the Agronomic Institute of Paraná - IAPAR, and Masasuke Mashima evaluated the different aspects related to their use in grapevines cultivated in subtropical regions (Kishino 1981), which stimulated other researchers to test their effectiveness in different grape growing regions.

The calcium cyanamide solution is prepared with hot water two to three hours before the application. Brushing is made easier if 5 mL to 10 mL adhesive spreader is added to 10 L of the solution. Reapplication is necessary if it rains in a period of 24 hours. For 'Niagara' and 'Isabella' grapevines, a 20% solution of calcium cyanamide is recommended.

In subsequent studies in the 1970s, it was found that during the process of preparation of calcium cyanamide solution, hydrogen cyanamide (H_2CN_2) was formed, a compound that has an effect as bud burst induction. Its synthesis was later developed by the chemical industry, and a few years after it was launched on the market a commercial product of easy preparation and classified as plant growth regulator, replacing in most cases the use of calcium cyanamide.

H_2CN_2 is the only molecule registered for bud burst in Brazil, commercialized under the name Dormex® (52% of active principle). During correlative dormancy, H_2CN_2 does not promote bud burst and is phytotoxic. In the endodormancy there is an increasing stimulation directly related to the dose. In the end of the endodormancy, a smaller concentration is required, and post-dormancy cyanamide is ineffective and may delay budding or be phytotoxic (Petri et al. 2006).

The mode of action of $H_2CN_2$is not yet fully understood, it may be related to its effects on the respiratory system of cells and by interference in some enzymatic processes, such as catalase activity (Shulman et al. al. 1986). The effect of oil, according to Erez et al. (1980), is due to the temporary anaerobic condition in the buds, resulting from the deprivation of oxygen by the oil cover, which causes "Pasteur´s effect" and consequently the ethanol production, substance responsible for overcoming dormancy. In spite of this effect, the use of mineral oil alone has a good effect on bud burst of apical buds, but its effect on bud burst of the lateral buds is lower to the association of mineral oil with H_2CN_2 (Petri et al. 2006).

H_2CN_2 shows rapid soil degradation in urea, nitrate and ammonia, as a nitrogen fertilizer, leaving no residues in the soil and plant (Pramanik et al. 2009; Goldbach et al. 1988). Studies indicate that the half-life of this product, after application, occurs in less than 2 days (Cai et al. 2012). In spite of the low persistence in the environment, this product is extremely caustic, causing irritation to the eyes and skin, hypotension, pulse acceleration, nausea, hot sensation and headaches in unprotected applicators. Ingestion of alcohol, before or after contact with the product, intensifies these symptoms. In view of this, the European Union has restricted the use of H_2CN_2, excluding this compound from Directives 91/414 / EEC in the year 2008 (Hernández and Craig 2011). Therefore, many viticulturists from other parts of the world are afraid that this product will be banned.

Its great advantage over calcium cyanamide is the ease in preparing the solution to be sprayed at concentrations ranging from 3% to 7% of the commercial product in tropical and subtropical regions. It should be applied within 72 hours after pruning, avoiding very hot hours or high winds. In case of rainfall up to two hours after application, it should be reapplied. Must not be associated with copper spray.

Due to its high toxicity, this compound must be kept away from food and children and in places with a temperature below 20°C, because above this temperature decomposition of the active ingredient with formation of crystalline precipitate in an irreversible process. It is not recommended consume alcoholic drinks the day before, during and one day after application, or as a precaution, until three days before and after. Contact with skin and eyes should be avoided as well as inhalation. The use of personal protective equipment during its preparation and application, such as gloves, goggles and masks, is mandatory. In case of poisoning, do not induce vomiting. No specific antidote is known.

2.2. Mineral and Vegetable Oils

The use of vegetable oils in agriculture has grown in recent years, especially with the launch of more products in the emulsifiable form. The information on the use of oils to induce sprouting of temperate climate fruits is restricted to mineral oils, with little information regarding the use of vegetable oils for this purpose (Hawerroth et al. 2010)

In a study carried out in the North of the State of Paraná, Brazil, a region with a subtropical climate, it was verified that the 'BRS Carmem' grapevine has a higher bud burst percentage with the use of vegetable oil, especially when associated with mineral oil (Figure 2).

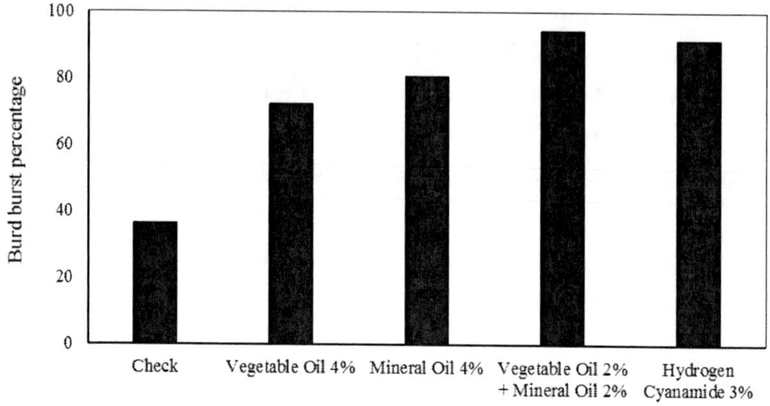

Figure 2. Percentage of bud burst of ‘BRS Carmem’ grapevine cultivated in Northern Paraná, Brazil and submitted to alternative treatments for induction of buds in subtropical climate. Londrina, PR, 2014.

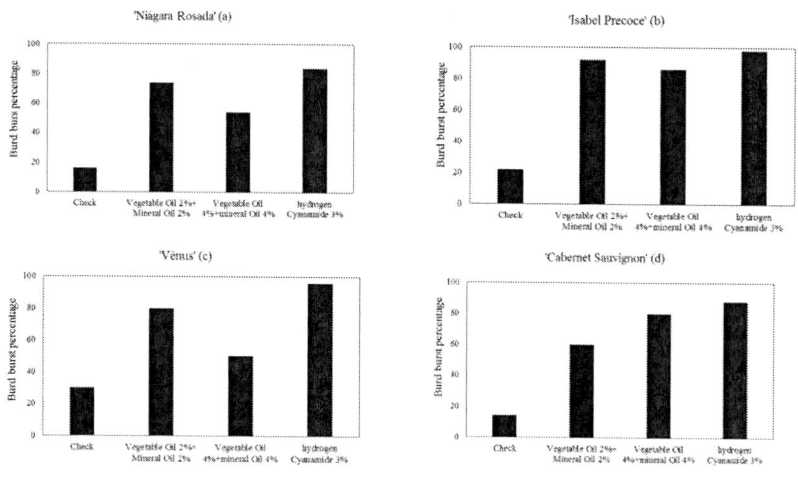

Figure 3. Percentage of bud burst of grapevines ‘Niagara Rosada’ (a), ‘Isabella Precoce’ (b), ‘Venus’ (c) and ‘Cabernet Sauvignon’ in the West of Paraná State, Brazil.

Similar results were obtained in the West of Paraná State, Brazil, where the treatments with vegetable oil associated to the mineral oil presented a high percentage of bud burst for the 'Niagara Rosada', 'Isabella Precoce', 'Venus' and 'Cabernet Sauvignon' (Figure 3). For the American and European cultivars, the best oils concentration was 2% and 4%, respectively.

The induction of bud burst occurs in several mechanisms, and some products directly affect bud respiration by the inhibition of oxidative phosphorylation, such as hydrogen cyanamide (Pérez and Lira 2005). Others affect respiration by creating anaerobic conditions, creating the "Pasteur´s effect", such as mineral and vegetable oil (Erez et al 1980). In a study carried out with 'BRS Carmem' (Table 2) it was verified that reduction of catalase activity in buds submitted to the application of mineral oil associated to the mineral oil is similar to the buds that received hydrogen cyanamide.

Table 1. Activity of catalase in grapevine buds 'BRS Carmem' zero and 24 hours after the application of treatments for artificial induction of budding in subtropical climate. Londrina, PR. 2014

Treatment	Catalase activity (μmol/min/mg protein)	
	0h	24h
Control	4.7017	4.7138
Vegetable oil 2%	4.5018	4.9698
Mineral oil 2%	3.3252	3.1443
Vegetable oil 2%+ Mineral oil 2%	4.0865	2.5086
Hydrogenated cyanamide 3%	3.9034	2.3497

2.3. Garlic Extract

Another compound with potential to promote bud burst is the garlic extract, because besides presenting potential for bud burst, it does not cause plant phytotoxicity (Kubota et al. 2000). The application of the garlic paste in the region of the pruning cut of 'Moscatel de Alexandria' vines shoots stimulates bud burst more effectively than the application of 20% solution of calcium cyanamide (Kubota e Miyamuki 1992).

'Cabernet Sauvignon' grapevines submitted to application of a commercial product based on garlic (Bioalho®) at 3% provides 70% of bud burst at 35 days after its application (Botelho et al. 2007). In general, the genus *Allium* presents promising results on bud burst. Garlic extract and chives at 10% associated with mineral oil improve significantly the bud burst of 'Cabernet Sauvignon' grapevines. The substances present in garlic responsible for dormancy release were identified as diallyl disulfide, the compound responsible for the aroma of garlic and other species of the genus *Allium* (Kubota et al. 1999).

2.4. Pruning and Bud Burst Induction Systems of Grapevine in Regions of Tropical and Subtropical Climates

In tropical and subtropical regions there is the possibility of obtaining multiple annual grape harvests, while in temperate regions only one harvest per year is possible, due to longer occurrence of low temperatures during fall-winter period, which makes the release of vines dormancy later. By performing a summer (or green) pruning, a late harvest can be obtained in the same vegetative cycle in tropical and subtropical regions. Among the main advantages of altering the cycle of vines are intensive cultivation, with more than one crop per year, where the supply of grapes is low or where climatic conditions are more favorable for fruit maturation, with lower rainfall.

In the North of the State of Paraná, Brazil, where climate is classified as Cfa (humid subtropical), the double annual harvest system is used for table grapes production. The first pruning is carried out at the end of winter, between July and early August, with subsequent application of hydrogen cyanamide 2.5-3.0% in two apical buds, and the grape harvest occurs between December and January. The second pruning (summer or green pruning) is carried out about one month after harvest, with subsequent application of hydrogen cyanamide 2.5-3.0% in apical buds, producing the off-season production in autumn, between May and June (Ricce; Caramori; Roberto, 2013). The applications of hydrogen cyanamide are carried out using a sponge attached to a wooden rod, so that the solution is applied only in two apical buds.

In the western region of the State of São Paulo, Brazil, where climate is classified as Aw (tropical semi-humid with dry winter), two pruning operations are performed to obtain a single production of table grapes. The cane pruning with subsequent application of hydrogenated cyanamide 3.0%, is carried out between April and May, and production occurs approximately 130 days after pruning, that is, during period from June to October, when the supply of table grapes on the market is low. Approximately one month after harvest, spur pruning is carried out with subsequent application of hydrogen cyanamide 3.0% in two basal buds. This pruning is very drastic and has the purpose of forming only new canes for next year's production, because practically no fruit is formed, since the fruitfullness of basal buds of these vines is low. As the region presents dry winter, the use of irrigation becomes necessary for the bunches' maturation period. The application of hydrogen cyanamide is similar to that described previously, spraying directly on buds.

In the south of the state of Minas Gerais, Brazil, where the climate is Cwb (humid temperate with dry winter), a similar technique is used for production of wine grapes, such as 'Syrah'. This practice consists in reversing the vine's cycle for the fall and winter period, in order to avoid the excesses of summer rains and high temperatures. Spur pruning is performed in mid-August, leaving two buds, and subsequent application of hydrogen cyanamide 2.5% in these buds. In this cycle, all inflorescences that arise are eliminated. From the month of January, when canes are already lignified, cane pruuning is carried out with subsequent application of hydrogen cyanamide 2.5%. In this cycle, the vines sprout about 10 to 12 days after pruning, the flowering occurs at the end of February, the bunches close at the end of March, and ripening begins in the mid to end of April, from June to early July, in which climatic conditions are more favorable to maturation of the bunches (Amorim, Favero, Regina 2005). Soon after, the vine rests around 30 days to start a new formation cycle.

In the Sub-Valley of the São Francisco River, Brazil, where the climate is Bsh (semi-arid hot) with average temperatures always high and rains are scarce, irrigation is required for the production of table grapes. Thus, in this region it is possible to obtain multiple annual crops at any time of the year. The traditional concept of winter chilling requirement for bud burst is completely countered in this region, with the combined use of tropical rootstocks such as 'IAC 313 Tropical', hydrogen cyanamide and irrigation management. Under these conditions, pruning made always one month after harvest are carried out at any time of the year alternating with medium and short canes (spurs) for renovation. After harvest, the irrigation is suspended for about a month in order to cause a water stress in the plant. After this period, irrigation is resumed with the application of 3.0-3.5% hydrogen cyanamide in two apical buds to induce a new productive cycle. The lowest doses of hydrogen cyanamide are recommended in the hottest periods of the year (September-April) and the highest in the milder months (May-August).

CONCLUSION

Viticulture, formerly restricted to well-defined winter areas where the grapevine cycle was also restricted by the imposition of low-temperature restrictive climatic conditions, spread to warmer subtropical and tropical climate regions with the development of technology, especially in relation to techniques for overcoming dormancy. It is possible in these areas to produce at different times and often completing more than one crop per year. The restrictions of the use of bud burst inductors due to environmental impact lead to the search for alternatives that cause less impact and may lead to the development of organic production. Research is still preliminary, but it already shows great potential for use, such as the use of plant extracts and vegetable oils.

REFERENCES

Amorim, D. A.; Favero, A. C.; Regina, M. A. (2005). Produção extemporânea da videira, cv. Syrah, nas condições do Sul de Minas Gerais [Extemporaneous production of the grapevine, cv. Syrah, in the southern conditions of Minas Gerais]. *Revista Brasileira de Fruticultura* 27(2): 327-331.

Botelho, R. V.; Pavanello, A. P.; Pires, E. J. P.; Terra, M. M.; Müller, M. M. L. (2007). Effects of chilling and garlic extract on bud dormancy release in Cabernet Sauvignon grapevine cuttings. *American Journal of Enology and Viticulture* 58(1): 402-404.

Campoy J. A.; Ruiz, D.; Egea, J. (2011). Dormancy in temperate fruit trees in a global warming context: a review. *Scientia Horticulturae* 130(1): 357-372.

El-Yazal, M. A. S.; El-Yazal, S. A. S.; Rady, M. M. (2014). Exogenous dormancy-breaking substances positively change endogenous phytohormones and amino acids during dormancy release in 'Anna' apple trees. *Plant Growth Regulation*, 72 (1): 211-220.

Erez, A.; Couvillon, G. A.; Kays, S. J. (1980). The effect of oxygen concentration on the release of peach leaf buds from rest. *Hort Science*, 15(1): 39-41.

Erez, A.; Yablowitz, Z.; Korcinski, R. (2000). Temperature and chemical effects on competing sinks in peach bud break. *Acta Horticulturae*, Brussels, 514(1): 51-58.

Gárcia, N. A.; Iribarne, C.; Palma, F.; Lluch, C. (2007). Inhibition of the catalase activity from Phaseolus vulgaris and Medicago sativa by sodium chloride. *Plant Physiology and Biochemistry,* 45(1): 535-541.

George, A. P., Broadley, R. H., Nissen, R. J.; Ward, G. Effects of new rest-breaking chemicals on flowering, shoot production and yield of subtropical tree crops (2002). *Acta Horticulturae*, 575(1): 835-840.

Gemma, H. Rest breaking in Delaware grape (1995). *Acta Horticulture*, 395(1): 127-133.

Hawerroth, F. J.; Petri, J. L.; Leite, G. B. (2010) Cianamida hidrogenada, óleos mineral e vegetal na brotação de gemas e produção de macieiras 'Royal Gala' [Hydrogen cyanamide, mineral and vegetable oils in the burd burst and 'Royal Gala' apple production]. *Semina: Ciências Agrárias*, 31(1): 1145-1154.

Horvath, D. P.; Anderson, J. V.; Chao, W. S.; Foley, M. E. Knowing when to grow: signals regulating bud dormancy (2003). *Trends in Plant Science*, 8(1): 534-540.

Kishino, A. Y.; Nascimento, E. C.; Tsuneta, M; Carvalho, S.D.C.; Mashima, M. (1978) Considerações sobre o uso de calciocianamida visando estimular a brotação da videira [Considerations on the use of calcium cyanocyanide to stimulate bud burst of the grapevine]. *Informe da Pesquisa 7 IAPAR* (1):1-7.

Kishino, A. Y. (1981). *Videira 'Itália' (Vitis vinifera* L.*): produção tardia da uva com variações no sistema e na época de poda* [*'Italian' grapevine (Vitis vinifera* L.*): Late Production of grape under different training and prunin seasons*]. Dissertação (Mestrado em Fitotecnia) - Escola Superior de Agricultura "Luiz de Queiroz," Universidade de São Paulo.

Kishino, A. Y.; Mashima, M. (1980) Uva. In: IAPAR. *Manual agropecuário para o Paraná* [*Agricultural Manual for Paraná*] 1(1)139-177.

Kishino, A. Y.; Genta, W.; Roberto, S. R. (2007). Introdução: produção de uva no Paraná [Introduction: grape production in Paraná]. In: Kishino, A. Y.; Carvalho, S. L. C. de; Roberto, S. R. *Viticultura tropical: o sistema de produção do Paraná*. 1(1): 22-23.

Kuroi, I. (1974). Studies on the growth promotion of grape vines by the lime nitrogen treatment during the rest period. *Memoirs of the Faculty of Agriculture*, Niigata, 12(1): 1-71.

Kubota, N.; Yasushi, Y.; Koji, T.; Kazuyoshi, K.; Tesuo, H.; SHOJI, N. (1999) Identification of active substances in garlic responsible for breaking bud dormancy in grapevines. *Journal of the Japanese Society for Horticultural Science*, 68 (1): 1111-1117.

Kubota, N.; Matthews, M. A.; Takahagi, T.; Kliewer, W. M. (2000). Effects of garlic preparations and of calcium and hydrogen cyanamides on budbreak of grapevines grown in greenhouses. *American Journal of Enology and Viticulture*, 51(1): 409-414.

Kubota, N.; Miyamuki, M. Breaking bud dormancy in grapevines with garlic paste (1992). *Journal of the American Society for Horticultural Science*, 117(1): 898-901.

Kuroi, I.; Shiraishi, S. I.; Mano, S. I. Studies on breaking the dormancy of grapevines. I. Effects of lime nitrogen treatment for shortening the rest period of glass house grown grapevine (1963). *Journal of the Japanese Society for Horticultural Science* 32(1): 175-80.

Lang, G. A; Early, J. A.; Martin, G. C.; Darnell, R. L. (1987) Endo, para, ecodormancy: Physiological terminology and classification for dormancy, esearch. *Hort Science*, 22(3): 371-377.

Lavee, S.; May, P. (1997). Dormancy of grapevines buds: facts and speculation. *Australian Journal of Grape and Wine Research*, 3(1): 31-46.

Malavolta, E. (1981). *Manual de química agrícola: adubos e adubação.* [*Manual of agricultural chemistry: fertilizers and fertilization*] 3(1): 596.

Mullins, M. G.; Bouquet, A. Williams, L. E. (1992) *Biology of the grapevine.* University Press.

Olsen, J. E. Mechanisms of dormancy regulation (2006). *Acta Horticulturae*, Saltillo, 727(1): 157-166.

Or, E.; Vilozny, I.; Eyal, Y.; Ogrodovitch, A. (2000) The transduction of the signal for grape bud dormancy breaking induced by hydrogen cyanamide may involve the SNF-like protein kinase GDBRPK. *Plant Molecular Biology,* 43(1): 483-494.

Pérez, F. J.; Lira, W. (2005) Possible role of catalase in post-dormancy bud break in grapevines. *Journal of Plant Physiology*, 162(1): 301-308.

Petri, J. L.; Herter, F. G. (2002). Dormência e indução a brotação [Dormancy and burd burst inducing]. In: Monteiro, L. B; De Mio, L. L. M.; Serrat, B. M.; Motta, A. C.; Cuquel, F. L. *Fruteiras de Caroço: Uma visão ecológica.* 1(1): 119-127.

Petri, J. L. (2006). Formação de flores, polinização e fertilização [Flowering, pollination and fertilization.]. In: *EPAGRI. A cultura da macieira.* 1(1): p.229-260.

Pinto, M.; Lira, W.; Ugalde, H.; Pérez, F. (2007) *Fisiologia de la latência de lãs yemas de vid: hipótesis actuales* [*Physiology of the latência of the yolks of vine: current hypothesis*]. Grupo de Investigación Enológica (GIE). Accessed in August 02, 2018: http://www.gie.uchile.cl/publicaciones/index.html.

Pires, E. J. P.; Pommer, C. V. In: Pommer, C. V. (2003) Uva Tecnologia de Produção, Pós-colheita, Mercado [Production Technology, Post-harvest, Market]. Porto Alegre: *Cinco Continentes*, 1(1): 250-294.

Powell, L. E. Hormonal aspects of bud and seed dormancy in temperate zone woody plants (1987). *Hort Science*, Alexandria, 22(1): 845-850.

Prasad, K. T.; Anderson, M. D.; Martin, B. A.; Stewart, C. R. (1994) Evidence for chilling-induced oxidadtive stress in maize seedlings and a regulatory role for hydrogen peroxide. *The Plant Cell*, Rockville, 6(1): 65.

Ricce, W. S.; Caramori, P. H.; Roberto, S. R. (2013) Potencial climático para a produção de uvas em sistema de dupla poda anual no estado do Paraná [Climatic potential for the production of grapes in an annual double pruning training in the state of Paraná]. *Bragantia*, 72(4): 408-415.

Ruiz, D.; Campoy, J. A.; Egea, J. (2007) Chilling and heat requirements of apricot cultivars for flowering. *Environmental and Experimental Botany*, 61 (1): 254-263.

Saure, M. C. (1985) Dormancy release in deciduous fruit trees. *Horticultural Reviews*, 7(1): 239-299.

In: Plant Dormancy
Editor: Renato V. Botelho

ISBN: 978-1-53615-380-4

Chapter 6

MANAGEMENT, PHYSIOLOGY AND GENETICS OF POTATO TUBER DORMANCY

Jackson Kawakami[1,*], Luiz Henrique Ilkiu Vidal[1] and Giovani Olegário da Silva[2]

[1]Department of Agronomy, Midwestern Parana State University - UNICENTRO, Guarapuava, Paraná State, Brazil

[2]Embrapa Vegetables - Brazilian Agricultural Research Corporation – EMBRAPA, Canoinhas, Santa Catarina State, Brazil

ABSTRACT

In a world with increasing population, food production to feed this population growth is crucial. Potato ranks third as the main source of food in the world. In potato's production in the field, the use of tuber seed with appropriated sprout growth is important to achieve high yields. The start of the sprout growth of potato tubers, in turn, is dependent on the breaking of dormancy. In addition, after tuber harvest, it is important to maintain tuber dormancy in order to keep the quality of this tuber. In this chapter, we stated the definition of potato tuber dormancy, showed the influence of dormancy on tuber yield, gave the main managements to keep or break

* Corresponding Author's E-mail: jkawakami@unicentro.br.

tuber dormancy and reported on the factors affecting dormancy period. In the sequence we discuss the physiological aspects of tuber dormancy, i.e., the main plant hormones that are related to the regulation of this process. Finally, we present the genetics of tuber dormancy and release, presenting the newest findings in this area of study and the future steps to achieve a better understanding of the genetic control of the breaking of dormancy in potato tuber. We hope that this text would help technicians and growers to better understand and manage this important food crop.

Keywords: *Solanum tuberosum*, sprout, plant regulators, storage.

1. INTRODUCTION

The potato tuber is a modified stem, usually grown below the soil surface. The tubers form at the end of stolon and start to accumulate carbohydrates. At the time when tubers are formed, they are dormant, i.e., they will not sprout. So, the term 'dormancy' in potato tubers is related as "the physiological state of the tuber in which autonomous sprout growth will not occur, even when placed under ideal natural conditions for sprouting (darkness, temperature 15-20 °C, relative humidity about 90%)" (Reust 1986). This kind of dormancy is also known as true state of dormancy, innate dormancy or *endodormancy* (Jefferies and Lawson 1991). Tubers can be kept dormant because they lack conditions to sprout. Usually, tubers are kept dormant by storing them at temperature around 2-4 °C. This kind of dormancy is known as enforced dormancy or *ecodormancy* (Suttle 2007). Finally, the dormancy can be broken but not all sprout will grow; usually just the apical sprout grows. This kind of dormancy is known as *paradormancy* (Suttle 2007).

The break of dormancy can be stated when 90% of a sample of 20 tubers at 20 °C for three weeks develop sprouts of at least 3 mm in lengths (Krijthe 1962). The measurement of dormancy period, however, varies in the literature. Usually, authors measure dormancy period from harvest until the break of dormancy. The time of harvest, however, can be carried out at different physiological stages: when plants still have green foliage and are destroyed, when plants have just senesced (naturally or killed by pests) or some period after plant senescence. In terms of plant physiology, the best dormancy period would be the time between tuber initiation and break of dormancy. Since field identification of tuber initiation is difficult, the time from planting or harvesting to break of dormancy should be preferably used as the dormancy period (Burton et al. 1992).

Storage can not be used to increase the quality of the tubers as a propagating material. If well stored, tubers can maintain their physiological and physical characteristics, but if not appropriately stored, tubers will lose these characteristics (Struik and Wiersema 2012).

2. Management

2.1. Impact of Dormancy on Potato Plant Growth and Tuber Yield

Depending on the physiological age, tubers have different sprout behaviors. Tubers that are allowed to break their endodormancy usually show apical dominance, i.e., there is one dominant sprout that has more vigorous growth than the others sprouts (young). As time pass by, apical dominance weakens, and others sprouts start to growth with the same vigor as the apical sprout (middle). After this stage if tuber is not planted and ages, small tubers can be formed directly at the sprouts, a phenomenon known as little tuber disorder (old) (Wiersema 1985).

The physiological age of tubers, young, middle or old, has impact on subsequent plant growth and tuber yield in the field. One important difference among plants originated from young, middle or old tubers is the number of stems that the tubers develop. In general, the younger the tuber, the lower the number of emerged stem (Struik and Wiersema 2001). Potato plants grown from young tubers usually produce lower number of tubers, but these tubers have higher average weight. Old tubers, however, originate plants with higher number of stems and higher number of small tubers. Also, this type of tubers generally originate plants with faster initial growth and shorter growth period (the period of days between planting and harvesting)

(Burton et al. 1992). If the plants are allowed to grow until the end of their life, the use of middle age tubers, without endodormancy and with no apical dominance (paradormancy) will result in higher tuber yield (Zaag and Loon 1987).

2.2. Breaking Dormancy

In countries located with temperate climate like EUA, Canada, European countries and Japan, the main objective in the management of tuber dormancy is to maintain tuber dormancy. This is because the window to potato planting is well definite and occurs once a year. Therefore, after harvest, tubers must be kept dormant for around half a year before being used as a seed tuber in the next crop year. On the other hand, countries located in lower latitude like India, Brazil and Bangladesh can grow more than one potato crop per year. In these countries, knowledge to break tuber dormancy is valuable. Also, dormancy breaking is useful for seed tuber for export, rapid post-harvest testing and seed multiplication program (Muthoni et al. 2014).

Among the products used to break potato seed dormancy, it can be cited:

- Rindite, a mixture composed by ethylene chlorohydrin, ethylene dichloride and carbon tetrachloride, in the proportions 7: 3: 1 (Coleman 1987). Rindite is used in the concentration of 1 L m^{-3} for 48 h (Es and Hartmans 1981). This substance is toxic and must be used with care. This product will be prohibited for commercial use in all countries of the world, so although effective to break tuber dormancy (Struik and Wiersema 2012), alternatives products must be developed.
- Gibberellic acid (GA). Usually GA is used as a solution (5 ppm or 5 mg L^{-1}) where cut tubers are immersed for 1-5 min (Struik and Wiersema 2012). Study concluded that dipping cut tubers in 10 to 100 ppm solution of GA_3 is effective to overcome tuber dormancy in 10 cultivars (Jansky and Hamernik 2014). The use of GA_3 with

other chemicals (thiourea and daminozide) was reported as a good approach to break dormancy (Wróbel et al. 2017). Usually, application of GA just after harvesting or, as soon as possible, after tuber cutting, is the most effective (Burton et al. 1992).

- Carbon disulphide (CS_2) is a gas used commonly in countries like Brazil. Usually growers use 12.5-30 mL m^{-3} for 3 days for around 500 kg of tuber seed to break dormancy. This gas is also toxic, inflammable and explosive and must be used with care.
- Thiourea, a catalase inhibitor, is another product with sulphur used to break dormancy in potato. Usually, this product is applied in cut potatoes in a dose of 1-2% thiourea solution for 60-120 minutes (Mani et al. 2014). This product is used when tubers are weakly dormant or when tubers are at the end of the dormancy period (Struik and Wiersema 2012).

In some cases, dormancy of tubers can be broken without the use of chemicals. Management of the temperature is one way to break tuber dormancy. Constant storage temperature of 28 °C, or limited period of 28 °C followed by a period of 18 °C can break dormancy (Struik and Wiersema 2012). The effectiveness of these temperature, however, depends on the cultivar. Another way to break tuber dormancy is to cut the seed. Tubers that were cut will produce GA, and this probably helps to break dormancy. The spread of diseases, like bacteria and virus, and the labor to cut the tubers, are the main disadvantages of this process.

2.3. Keeping Dormancy

When potatoes are sold in fresh marked, it is not appropriated that potatoes start to sprout. This is because tuber quality decreases with sprouting (Mani et al. 2014). Also, in the potato seed market, controlling the time of sprouting initiation is important. Usually, tubers are kept dormant by low temperature (Burton et al. 1992). Constant temperatures of 2-4 °C is efficient to maintain tuber dormancy (ecodormancy).

Among the products usually used to keep tuber dormancy, we can cite:

- Dimethylnaphthalenes (DMN). There are several isomers of this product and 5-30% concentration in synthetic mixture are commonly used (Burton et al. 1992).
- Isopropyl N-phenlcarbamate (propham, IPC) and isopropyl N-(3-chloropheny) carbamate (chlorphopham, CIPC) are applied as a mixture to inhibit sprout growth (Burton et al. 1992). Usually the rate of 10 g of the products t^{-1} of tubers are used. CIPC is the synthetic sprout inhibitor used in many countries; and the maximum CIPC residue level allowed is between 10-30 mg kg^{-1} (Bethke 2014).

3. Factors Affecting the Dormancy Period

3.1. Genetic

Generally, wild potato species have long dormancy period while cultivated potatoes have short dormancy, but there are exceptions (Suttle 2007). The tubers of wild potato *Solanum jamesii* can maintain the ability to sprout after eight year stored at 4 °C (Bamberg 2010). There is difference in this period among cultivars (Danieli et al. 2018). For example, in an experiment with 10 cultivars stored at 10 °C, Russet Burbank was the cultivar with longest dormancy: 123 days (between harvest and the end of dormancy), while Darkchip was the shortest: 64 days (Bogucki and Nelson 1980). In this same study, the dormant period of the other cultivars varied from 74 to 86 days.

Apparently, there is no relation between dormancy period and the maturity class of the cultivars (Burton et al. 1992).

Tuber Size

In genneral larger tubers have shorter dormancy than smaller ones, because they are phisiologicaly older. This characteristic, however, varies with cultivar (Ittersum 1992).

3.2. Environment

Temperature

Temperature during storage has high influence on the length of dormancy. With temperature between 3-25 °C, the higher the temperature, the shorter the length of dormancy (Burton 1989). Tubers stored at 3 °C will delay the sprout growth, these tubers will be in ecodormancy (Muthoni et al. 2014), although this temperature varies depending on the cultivar (Struik and Wiersema 2012). Also, the variation in the temperature during storage promotes the break of dormancy. Tubers kept at constant 15 °C will sprout later than tubers kept at 2 °C for two weeks and then moved to 15 °C (Thomas and Wurr 1976). Temperature of 28 °C seems to result in maximum speed of dormancy release (Struik and Wiersema 2012). Temperatures around 18 °C promotes the sprout growth (Struik and Wiersema 2012).

Normally, the higher the temperature during plant growth, the shorter the dormancy of harvested tubers (Muthoni et al. 2014). Some authors, however, stated that field temperature has low influence on tuber dormancy (Danieli et al. 2018). Temperature variation during the day also has influence on tuber dormancy: the higher the variation, the shorter the dormancy (Ittersum and Scholte 1992).

Light

In immature tubers, light may shorten the dormancy period compared with darkness, but in mature tubers the effect is the opposite (Struik and Wiersema 2012). Light can also contribute to maintain dormancy (ecodormancy) especially with blue (500 nm) and red / far-red (650 nm) wavelengths (McGee et al. 1987).

Atmosphere

The relative humidity of the air has influence on dormancy. High air relative humidity favors dormancy release (Struik and Wiersema 2012). This effect, however, is lower than that caused by other factors like temperature and chemicals.

Other atmosphere elements, like CO_2, O_2, have low influence on endodormancy, but high influence on ecodormancy, i.e., the sprout growth (Struik and Wiersema 2012).

4. Physiological Aspects of Potato Tuber Dormacy

Among the main substances related with the start, maintenance and overcome of dormancy, we can highlight the plant hormones. Suttle (2007) states that since the end of 1940 decade, researchers already knew the plant growth inhibitory substances acting in tuber dormancy.

The classical group of plant hormones is formed by 5 substances groups: auxin, gibberellins, cytokinin, abscisic acid and ethylene. However, recently other substances have been studied due to their important metabolic functions in the plants growth and development. Among these new substances we can highlight the brassinosteroids and jasmonates.

The technical advances in chromatography and analytical methodology of research has allowed along the last decades precise evaluations in the quantification and function made by the endogenous plant hormones in potato tuber dormancy (Suttle 2004).

In the sequence, it will be discussed the main role of each group or plant hormones related with potato tuber dormancy.

4.1. Abscisic Acid

Among the distinct groups of plant hormones, the abscisic acid (ABA) is considered a growth inhibitor. One of its main and more pronounced physiological effect, both in seeds and in plant sprouts, is the induction and maintenance of dormancy.

At harvest, ABA concentration in tubers is generally high and decreases during the storage time. This variation in ABA concentration coincide with dormancy overcoming. Therefore, ABA content must decline at large extent before the induction of sprouts growth (Muthoni et al. 2014).

Chronological analysis in the endogenous concentration changes of this hormone shows evidences of the function made by ABA in the tuber dormancy maintenance (Destefano-Beltran et al. 2006). These authors suggest that the ABA synthesis for tuber dormancy maintenance is the result of the dynamic balance between the synthesis and the gradual degradation of ABA during tuber development.

The identification of the main ABA synthesis sites in tubers tissues is very important to the comprehension of the relation between this hormone metabolism and the tubers growth. Some researches indicate that tuber meristematic cells are both target cells as well as the main synthesis locations of ABA (Nambara and Marion-Poll 2005, Destefano-Beltran et al. 2006).

4.2. Ethylene

This plant hormone has among its main physiological effects the induction of tissues senescence. It is directly related to fruit ripeness and acts in the leaves and fruits abscission, among others important roles in the plant.

According to Suttle (2004), from middle decade of 1920 there were already reports that ethylene could influence breaking the dormancy in potato tubers. However, the function of this hormone may vary according to the concentration and exposition time.

The involvement of ethylene in the endodormancy regulation is still uncertain. This hormone favors the initiation and the maintenance of tubers dormancy, but these processes are still not completely elucidated (Aksenova et al. 2013). Some possible mechanisms could be the interaction between ethylene and ABA in tuber dormancy regulation; and ABA may act as a mediator of stages enhancement of ethylene dependent dormancy.

4.3. Gibberellins

This plant hormone plays many physiological effects of large importance in plants growth and development. It has fundamental function in stem elongation, and in the release of dormance in seeds and sprouts. In many times it has opposite effect to ABA.

Suttle (2004) says that since mid-1950 studies confirm that exogenous spraying of gibberellins (GAs) may overcome potato tuber dormancy, and exogenous GAs are not closely related to the dormancy control, but they have a function on growth and elongation sequence in the sprouts.

The GAs are a group formed by more than 120 different analogous substances, but just few of them are physiological active form of this hormone; the majority are just intermediate metabolites (Taiz et al. 2017). Among the active GAs, GA_3 highlights by its effect on dormancy release and sprouting of minitubers of different potato cultivars. However, the exogenous spraying of high concentration of GA_3 may initiates adverse effects in plants, like exaggerated stem growth, reduction on roots development, delaying tuber formation, leaves compaction and tubers malformation (Mani et al. 2014).

It seems that a prerequisite for dormancy breaking by GAs is translocation of this hormone to the tuber sprouts (Alexopoulos et al. 2008). Hypothetically the GAs regulate reserve mobilization through changes in intracellular compartmentalization (Coleman 1987).

Apparently, the endogenous activities of GAs are low during the tuber dormancy and increase after the sprout growth and during the tubers storage; the dormancy may be broken using GAs treatments (Akesenova et al. 2013).

4.4. Cytokinin

Cytokinin (CK) is a group of plant hormones that is directly related with cell division, among others metabolic functions. Experiments of 1970 decade demonstrated that both natural and synthetic CK can break tuber dormancy. Coleman (1987) reports that exogenous CK are effective in breaking tuber dormancy just during short periods immediately after the dormancy development and immediately before the end of dormancy.

After harvest, tubers are in endodormancy, showing low CK concentration and do not respond to sprout induction treatments. During storage as time pass by the dormant conditions "weakens", the CK concentrations remain low, but the tubers start to positively respond to dormancy breaking treatments, including exogenous application of hormones (Suttle et al. 2014). Dormant tubers show time-dependent increase in CK sensibility. This increase in CK sensibility is not accompanied by CK metabolism changes, suggesting that CK receptor and/or signal transduction pathways are affected by dormancy status (Suttle 2001).

4.5. Auxins

Auxins are the first group of plant hormones to have their physiological action described and tested in plant tissues. Among the principal effects in plant tissues, we can highlight cellular enlargement and the rooting capacity. Auxins can act in the maintenance of tissues juvenility, vascular bundle differentiation, among other important functions of plant development (Taiz et al. 2017).

The first studies demonstrating the effect of free auxin and esterified auxin on sprouting inhibition of dormant potato tubers are from the beginning of 1940 decade (Suttle 2001). The results from these first studies suggest that endogenous auxins are not directly related with tuber dormancy regulation, but with subsequent sprouting regulation of tubers (Suttle 2007).

Suknova et al. (1993), using high-performance liquid chromatography (HPLC) equipment coupled with fluorometric detection, observed that the auxin content remains constant during potato tuber dormancy, increasing just after the beginning of sprout growth. These observations suggest that auxin concentration reflect more the growth rate of sprouts than the tuber dormancy.

4.6. Jasmonic Acid

Jasmonates (JAs) and their correlate biosynthetic compounds (oxylipins) have significant effect on potato plant development. The tuberonic acid, a hydroxylated derivative of JA, is one of the most potent in vitro tuberization inductor, and may have important roles in plant tuberization (Yoshihara et al. 1989). There was already related effect of sprouting inhibition of tubers during storage, but its function as a regulator hormone of dormancy is still unclear (Suttle 2007).

Data from different researches with JA functioning dynamic during tuber dormancy and sprouting are contradictory and the effects of JAs treatment on tubers are also ambiguous. More researches with this class of hormones are necessary to elucidate the JA involvement in the tubers dormancy and sprouting process (Aksenova et al. 2013).

In Table 1, it is outlined the summary of hormonal regulation of tuber dormancy and sprouting.

Although tuber dormancy has been extensively studied during decades, the mechanisms that control all these processes are still poorly understood. For example, Viola et al. (2007) highlights that it is still not clear if the status of sprout growth is controlled by the sprout itself (endodormancy) or by the rest of the tubers (paradormency).

Many of the data in the literature are fragmented and contradictory. This confusion is due to different analysis methods of tested hormones (bioassay x chemical-physical analysis) and by choice of the tissues that is analyzed (Suttle 2007). As natural dormancy happens through a variable period of tuber storage, physiological changes not directly related with dormancy also occurs in the tubers; this may cause some misunderstood of data interpretation (Suttle 2014).

Table 1. Summary of hormonal regulation of potato tuber dormancy and sprouting

Phases of tuber dormancy and sprouting: stimulating phytohormones	Physiological processes			
	Buds		Tuber tissues	
	Growth characteristics	Metabolism specific features	Growth characteristics	Metabolism specific features
Dormancy: ABA, Ethylene,	Growth arrest; Blocking G1/S-phase transition	Low level of metabolism; Limited supply with substrates	Absence of growth and morphogenesis	Low level of metabolism; preservation of storage carbohydrates and patatins
Sprouting: IAA, CK, GA	Onset of growth and morphogenesis; Removal of blocking G1/S-phase transition	Activation of metabolism; Enhanced supply with substrates	Absence of growth and morphogenesis	Transition from storage to source function; Enhanced activity of carbohydrate metabolism
Sprout growth: CK, GA	Active growth	High level of metabolism; active supply with substrates	Absence of growth and morphogenesis	Active supply of seedlings with energetic and structural materials

Adapted from Aksenova et al. (2013).

Potato tuber dormancy is influenced in a complex way by both genetic and environmental factors, mediated by hormonal status changes (Mani et al. 2014). In the past, many researches focused on specific features of dormancy control, like the hormones, or the cellular cycle regulation. Nowadays studies use the quantitative genetic power to uncover traits by quantitative traits loci (QTL) analysis. This has reveled many genetic loci involved in dormancy control.

5. The Genetic of Tuber Dormancy Control and Release

Potato crop (*Solanum tuberosum* L.) is a tetraploid crop (2n = 4 x 12 = 48 chromosomes) and has multiple alleles controlling some characteristics.

These characteristics result in a complex genetic control. In tuber dormancy the high hormonal interaction adds more complexity to this process. Moreover, there are many genes involved in the tuber dormancy and therefore the character is considered as quantitative.

A high variation in the dormancy period has been observed in cultivated potato and in the related wild species (Muthoni et al. 2014). In general, long dormant period are found in wild species of potato. On the other hand, short dormant period is observed in cultivated species (Suttle 2007). An exception of this pattern is the group *Solanum phureja*, a wild species that, in general, is characterized by short tuber and seed dormancy (Huamân and Spooner 2002).

Simmonds (1964), studying diallelic crossing scheme with six cultivated diploid entries of subspecies *Solanum tuberosum* spp. andigena, verified that some entries transmitted efficiently their characteristics of longer or shorter dormancy. This author also observed that crossing with reciprocals showed significant parental effect, that is, different responses when determined entries were used as male or female in the crossings. Also, this author found polygenic addictive effect in the control of the character. Moreover, in almost all potato species, there is dormancy in the true seeds and in the tuber seeds, but in some diploid potatoes from South American there is higher variability in dormancy. Possibly this higher variability was caused by the human selection for clones adapted to cultivation in cold or hot places. This modification in both diploid and tetraploid potatoes was possible due to the existence of polygenic genetic variation and this modification was followed by the reduction of genetic variation in these populations, and in the development of the potato in the Europe (*Solanum tuberosum* spp. tuberosum). This selection favored higher dormancy, because the tubers had to be stored during winter times.

Although the basic mechanisms of genetic control of potato tuber dormancy are still poorly known, research information leads to the understanding of this process, and strengthen the concept of complex genetic control of this character. Therefore, studies with quantitative trait loci (QTL) shows that potato tuber dormancy is controlled by at least nine different loci (Van den Berg et al. 1996, Ewing et al. 2004).

Ewing et al. (2004), using a population derived from the crossing between cultivated specie *Solanum tuberosum* spp. tuberosum and the diploid specie *Solanum berthaultii* backcrossed with haploid *Solanum tuberosum* spp. tuberosum, found eight QTLs related with tuber dormancy. *S. berthaultii* is a specie of Bolivian's origin marked by short days tuberization, with long dormancy period, and resistance against a wide range of insect pests. Three of this QTLs are related with previously loci associated with sensibility to photoperiod, suggesting that these could represent genes that affect the hormone content of the tuber (Ewing et al. 2004). Two of these QTLs are related with loci associated with abscisic acid content (ABA) (Šimko et al. 1997).

In the same way, Van den Berg et al. (1996), using two segregating progeny of *Solanum tuberosum* with *S. berthaultii*, detected QTLs in nine chromosomes affecting tuber dormancy, isolated or with epistatic interaction. The wild species allele, found in the chromosome 2, contributed with high effect for longer dormancy. In the backcrossing with *S. berthaultii*, these authors verified that long dormancy seems to be a recessive character, and that addictive effect of QTLs contributes with 48% of the measured phenotypic variance; in addition, epistatic effect explain just another 4% of this value. On the other hand, in the backcrossing with *S. tuberosum*, addictive effect explained just 16% of the variance, while additional 24% was explained with the inclusion of epistatic effects. In this study, the QTL located in chromosome 2 (TG234), contributed with 31% of the variance. The presence of the allele T in this QTL leaded to the reduction of 29 days in the dormancy period in the evaluated population. Three other QTLs, TG66 in the chromosome 3, TG443 in the chromosome 4, and TG41 in the chromosome 8, showed lower effects. However, in all three the presence of allele T lowed the dormancy.

In another study with QTL, Freyre et al. (1994) evaluated a population of 110 plants, originated from a crossing between female haploid plants of *S. tuberosum* (2x) crossed with a wild specie *S. chacoense*, and a male parent of *S. phureja* specie. The population was characterized with 10 isoenzyme loci, 44 restriction fragment length polymorphisms of DNA (RFLPs) and 63 random amplified polymorphic DNAs (RAPDs). The authors used 87 of the female parent segregating loci to construct a linkage map with 10 of the 12 chromosomes in the genome. Dormancy, measured as number of days to sprouting after harvest, varied from 10 to 90 days, with average of 19 days. QTLs were mapped by analysis of variance to verify the association of each marked locus with the clones dormancy. It was identified six QTLs located in the chromosome 2, 3, 4, 5, 7 and 8, each. The QTL with highest effect in dormancy was detected in chromosome 7. The authors developed a multilocus model with the locus with highest coefficient of determination in each QTL. This model explained 57.5% of the dormancy phenotypic variation. Analysis of variance showed that 7% of the possible epistatic interactions among markers were significant. When these were included in the main effect model, it explained 72.1% of dormancy phenotypic variation.

Although QTL analysis helps in the genetic control knowledge, there is variation according to the genetic materials used in the study and also with the ambient condition in which the tubers are placed, and the association of their phenotype with the molecular markers. In addition, despite of the high number of loci related with carbohydrate synthesis is already mapped in potato, future studies will be necessary to investigate candidate genes co-located with QTL associated with tuber formation and dormancy (Eck 2007).

Similar to the studies with QTLs, differential expression of transcript analysis in different plant tissues also helps to clarify the genetic control of tuber dormancy breaking. Studying genetic expression changes related to the start of sprouting of minitubers from nodal cuttings in hydroponics and using finguerprint with amplified fragment length polymorphism (AFLP) analysis, Bachem et al. (2000) estimated that 40,000 genes were expressed during all the physiological cycle of microtubers, since their formation until the sprouting. The genetic expression was higher during the start of tuberization, was lower during the dormant period, and increased again during sprouting. A total of 1,300 transcript derived cDNA fragments (TDFs) were identified during the dormancy breaking progression; of this, 13% showed changes in the expression during this period, with 4% of these genes lowering this expression during the initial stage of the dormancy, and 3% showing increase with sprouting. Most of these genes did not show changing during the sprouting or showed an irregular pattern of expression not related with physiological state. With the isolation and sequencing of 75 TDFs that changed the expression related with dormancy, it was verified that 64% of this had no similarity with any genes in the studied databases, but many were similar to cDNA sequence in database of expressed genes sequence (EST).

In another study of EST sequencing with different potato tissues, Ronning et al. (2003) identified almost 20,000 express sequences, from which 44% were similar to known genes. From this, approximately 2,200 specific of sprouting and 851 of dormant tuber skin (excluding sprout and periderm) were identified. Almost 50 genes were expressed stronger in the dormant microtubers skin than in microtubers at all, and more than 40 genes were expressed stronger in the eye with sprouting than in the dormant tubers skin. In similar study, Faivre-Rampant et al. (2004a) identified 385 ESTs; from this, 12, 9 and 3% were similar to ribosomal protein genes, stress-related genes, and transcription factor genes, respectively. One gene encoding for a member of auxin-response factor family (StARF6) was identified and analyzed more precisely, and its expression was higher in tuber meristem during early stage of sprouting. In the same way, Faivre-Rampant et al. (2004b) identified other factor of potato transcription (StTCP) that had an inverse expression pattern: its expression was higher in dormant meristem and was not detected in sprout tissue.

There is high variability in potato crop and mainly in the wild specie for potato tuber dormancy. Therefore, the plant breeding may evolve in the development of cultivars with target dormancy level in accordance with the adaptation necessity to different climate and management. However, the genetic control and consequently the molecular markers identification associated with this characteristic remains to be studied. Collectively, QTL and gene differential expression studies, as well as the use of new techniques like genetic silencing, may allow in the future a better understanding of the genetic control of the mechanisms related to dormancy release in potato tubers. Also, the use, for example, of the assisted selection for introgression of targeted genes from wild species or the use of genetic engineering to modify these characteristics may be useful to better understand dormancy in potato.

REFERENCES

Aksenova, N., Sergeeva, L., Konstantinova, T., Golyanovskaya, S., Kolachevskaya, O. and Romanov, G. 2013. Regulation of potato tuber dormancy and sprouting. *Russian Journal of Plant Physiology* 60 (3): 301-312.

Alexopoulos, A. A., Aivalakis, G., Akoumianakis, K. A. and Passam, H. C. 2008. Effect of gibberellic acid on the duration of dormancy of potato tubers produced by plants derived from true potato seed. *Postharvest Biology and Technology* 49:424-430.

Bachem, C., Van der Hoeven, R., Lucker, J., Oomen, R., Casarini, E. Jacobsen, E. and Visser, R. 2000. Functional genomic analysis of potato tuber life-cycle. *Potato Research* 43: 297-312.

Bamberg, J. B. 2010. Tuber dormancy lasting eight years in the wild potato Solanum jamesii. *American Journal of Potato Research* 87: 226- 228.

Bethke, P. C. 2014. Postharvest storage and physiology. In: *The potato: botany, production and uses,* edited by Navarre, R. and M. J. Pavek, 255-271. Oxfordshire: CABI.

Bogucki, S. and D. C. Nelson. 1980. Length of dormancy and sprouting characteristics of ten potato cultivars. *American Potato Journal* 57 (4): 151–57. doi:10.1007/BF02853866.

Coleman, W. K. 1987. Dormancy release in potato tubers: a review. *American Journal of Potato Research* 64 (2): 57–68.

Danieli, R., Blank, L., Salam, B. B., Malka, S. K., Teper-Bamnolker, P., Daus, A., Zig, U. Amichay, M., Shemer, Z., Gal-On, A. and Eshel, D. 2018. Postharvest temperature has a Greater Impact on Apical Dominance of Potato Seed-Tuber than field growing-degree days exposure. *Field Crops Research* 223: 105–12. doi:10.1016/ j.fcr.2018.03.020.

Destefano-Beltrán, L., Knauber, D., Huckle, L. and Suttle, J. 2006 Chemically forced dormancy termination mimics natural dormancy progression in potato tuber meristems by reducing ABA content and modifying expression of genes involved in regulating ABA synthesis and metabolism. *Journal of Experimental Botany* 57: 2879–2886.

Eck, H. J. van. Genetics of morphological and tuber traits, p.91-116. In: Vreugdenhil, D. *Potato biology and biotechnology: Advances and Perspectives*. 2007, Elselvier, 801p.

Ewing, E. E., Simko, I., Omer, E. A. and Davies, P. J. 2004. Polygene mapping as a tool to study the physiology of potato tuberization and dormancy. *American Journal of Potato Research* 81: 281-289.

Faivre-Rampant O., Cardle, L., Marshall, D., Viola, R. and Taylor, M. A. 2004a, Changes in gene expression during meristem activation process in Solaum tuberosum with a locus on the regulation of an auxin response factor gene. *Journal of Experimental Botany*, 55: 613-622.

Faivre-Rampant, O., Bryan, G. J., Roberts, A. G., Milbourne, D., Viola, R. and Taylor, M. A. 2004b. Regulated expression of a novel TCP domain transcription factor indicates an involvement in the control of meristem activation processes in Solanum tuberosum. *Journal of Experimental Botany*, 55: 951-953.

Freyre R., Warnke, S., Sosinski, B. and Douches, D. S. 1994. Quantitative trait locus analysis of tuber dormancy in diploid potato (Solanum spp.). *Theoretical and Applied Genetics* 89:474–480.

Huamán, Z. and Spooner, D. M. 2002. Reclassification of landrace populations of cultivated potatoes (Solanum sect. Petota). *American Journal of Botany* 86 (6): 947-965.

Ittersum, M. K. van and Scholte, K. 1992. Relation between growth conditions and dormancy of seed potatoes. 2. Effects of temperature. *Potato Research* 35: 365-375.

Jansky, S. and Hamernik, A. 2014. Rapid cycling of potato tuber generations by overcoming dormancy. *American Journal of Potato Research* 92: 148–52. doi:10.1007/s12230-014-9415-7.

Jefferies, R. A. and Lawson, H. M. 1991. A key for the stages of development of potato (Solanum tuberosum). *Annals of Applied Biology* 119:387-399.

Mani, F, Bettaieb, T., Doudech, N. and Hannachi, C. 2014. Physiological mechanisms for potato dormancy release and sprouting: A review. *African Crop Science Journal* 22 (2): 155–74.

Muthoni, J., Kabira, J., Shimelis, H., Melis, R. 2014. Regulation of potato tuber dormancy: A review. *Australian Journal of Crop Science* 8 (5): 754-759.

Nambara, E. and Marion-Poll, A. 2005. Abscisic acid biosynthesis and catabolism. *Annual Review of Plant Biology* 56: 165-185.

Reust, W. 1986. EAPR working group 'physiological age of the potato'. *Potato Research* 29:268-271.

Ronning, C. M., Stegalkina, S. S., Ascenzi, R. A., Bougri, O., Hart, A. L., Utterbach, T. R., Vanaken, S. E., Riedmuller, S. B., White, J. A., Cho, J., Pertea, G. M., Lee, Y., Karamycheva, S., Sultana, R., Tsai, J. Quakenbush, J., Giffiths, H. M., Retrepo, S., Smart, C. D., Fry, W. E., van der Hoeven, R., Tanksley, S., Zhang, P., Jin, H., Yamamoto, M. L. Baker, B. J. and Buell, C. R. 2003. Comparative analyses of potato expressed sequence tag libraries. *Plant Physiology* 131: 419-429.

Šimko I., McMurry, S., Yang, H., Manschot, A., Davies, P. J. and Ewing, E. E. Evidence from polygene mapping for a causal relationship between potato tuber dormancy and abscisic acid content. 1997. *Plant Physiology* 115: 1453-1459.

Simonds, N. W. 1964. The genetics of seed and tuber dormancy in the cultivated potatoes. *Heredity* 19: 489–504.

Sukhoval, L. S., Machackova, I., Eder, J., Bibik, N. D. and Korableva, N. P. 1993. Changes in the levels of free IAA and cytokinins in potato tubers during dormancy and sprouting. *Biologia Plantarum*. 35 (3): 387-391.

Suttle, J. C. 2001. Dormancy-related changes in cytokinin efficacy and metabolism in potato tubers during postharvest storage. *Plant Growth Regulation*135:199-206.

Suttle, J. C. 2004. Physiological regulation of potato tuber dormancy. *American Journal of Potato Research* 81: 253-262.

Suttle, J. C. 2007. Dormancy and sprouting. In: *Potato biology and biotechnology: advances and perspectives*, edited by Vreugdenhil, D., 287–309. Oxford: Elsevier.

Suttle, J. C., Huckle, L. L., Lu, S. and Knauber, D. C. 2014. Potato tuber cytokinin oxidase/dehydrogenase genes: Biochemical properties, activity, and expression during tuber dormancy progression. *Journal of Plant Physiology* 171: 448–457.

Taiz, L., Zeiger, E., Moller, I. M. and Murphy, A. *Fisiologia e desenvolvimento vegetal* [*Physiology and plant development*]. 6. ed. Porto Alegre-RS: Artmed, 2017. 888 p.

Thomas, T. H.; Wurr, D. C. E. 1976. Gibberellin and growth inhibitor changes in potato tuber buds in response to cold treatment. *Annals of Applied Biology*. 83: 317-320.

Van Der Berg, J. H., Ewing, E. E., Plaisted, R. L., McMurray, S. and Bonierbale, M. W. 1996. QTL analysis of potato tuber dormancy. *Theoretical and Applied Genetics* 93: 317-324.

Viola, R., Pelloux, J., van der Ploeg, A., Gillespie, T., Marquis, N., Roberts, A. G. and Hancock, R. D. 2007. Symplastic connection is required for bud outgrowth following dormancy in potato (*Solanum tuberosum* L.) tubers. *Plant, Cell & Environment* 30: 973–983.

Wiersema, S. G. Physiological development of potato seed tubers. Lima: International Potato Center, 1985. *Technical Information Bulletin 20*.

Wróbel, S., Kęsy, J. and Treder, K. 2017. Effect of Growth Regulators and Ethanol on Termination of Dormancy in Potato Tubers. *American Journal of Potato Research* 94: 544–55. doi:10.1007/s12230-017-9592-2.

Yoshihara T., Omer A., Koshino H., Sakamura S., Kikuta, Y. and Koda, Y. 1989. Structure of a tuber-inducing stimulus from potato leaves (Solanum tuberosum L.). *Agricultural and Biological Chemistry* 53: (10) 2835-2837.

Zaag, D. E. van der and Loon, C. D. van. 1987. Effect of physiological age on growth vigour of seed potatoes of two cultivars. 5. Review of literature and integration of some experimental results. *Potato Research* 30: 451-472.

In: Plant Dormancy
Editor: Renato V. Botelho
ISBN: 978-1-53615-380-4

Chapter 7

DORMANCY IN SEEDS AND PROPAGULES OF VEGETABLES

Cristhiano Kopanski Camargo[1,*], Juliano Tadeu Vilela de Resende[2], André Ricardo Zeist[3], André Gabriel[1] and Ricardo Antonio Zeist[1]

[1]Department of Agronomy, Midwestern Parana State University, Guarapuava, Paraná State, Brazil
[2]Department of Agronomy, Londrina Parana State University, Londrina, Paraná State, Brazil
[3]Department of Agronomy, West Paulista University, Presidente Prudente, São Paulo State, Brazil

ABSTRACT

Of the US $ 4.6 billion in world trade of vegetable seeds, Europe and Asia account for 70%. Of this amount, Brazil is responsible for moving about US $ 100 million. Considered by several experts as an excellent

* Corresponding Author's E-mail: kopanski1976@hotmail.com

business environment in the world context of the seed industry, Brazil manages to combine its agricultural vocation with a good technological system. Furthermore, the Country has a continental characteristic, providing suitable climatic conditions for the production of high quality seeds of tropical and temperate species. Among the difficulties encountered in the seed production sector, dormancy stands out, which is usually an undesirable characteristic in the olericulture, since the rapid germination and growth are required as a primordial factor. The knowledge of dormancy mechanisms and their duration in the different vegetable seeds are ecologically and economically important, since they help in the selection of specific treatments that act on the metabolism of the seed, releasing the embryo for development or making it apt to germinate. Dormancy allows distribution of germination time, and is also an adaptive feature that ensures the survival and perpetuation of the species in the different ecosystems. The germinative delay maintains the species at the seed stage, i.e., at the stage of the plant cycle in which its resistance to the unfavorable conditions is greater. Seed dormancy is controlled by the genotype associated with: environmental factors, such as temperature; physical factors, such as tegument thickness; and regulation of hormones such as abscisic acid and gibberellin. Generally, plants produce seeds with various levels of dormancy. This aspect helps to distribute their descendants through time and space. In addition, it allows the seeds not to germinate during the adverse periods, guaranteeing a time required for the dispersion to the greater geographic distance, providing the survival of the species. In order to standardize and potentiate the germination of some seeds, it is necessary to identify the type of dormancy. Thus, this chapter presents several specific methods for breaking dormancy in vegetable seeds, considering mainly aspects of physiological and morpho-anatomic of the seeds.

Keywords: propagation, olericulture, germination, dormancy release.

1. Introduction

The world market for vegetable seeds is estimated to be around US $ 4.6 billion, made up of the chain of research companies, international traders, multipliers, distributors and resellers. More than 70% of the commercialization of vegetable seeds takes place in Asia and Europe (ABCSEM 2016).

Many experts consider Brazil as an excellent business environment in the world context of the seed industry, because it aligns its agricultural vocation, a good technological system, and its continental characteristic, which favors the production of high quality seeds of different species. In 2016, the sector of vegetables seeds handled around R $ 1.5 billion, especially tomato, onion, carrot, melon, watermelon and lettuce seeds, which represent 60% of this market (Camargo Filho and Camargo 2017).

With the process of domestication of plants and through selections made by genetic breeding, currently most species or cultivars are practically free of the mechanisms of dormancy. The dormancy is usually an undesirable characteristic in the horticulture, where rapid germination and growth are required (Tedesco et al. 2001) and also for nurserymen, whose dormancy generates problems such as unevenness among the seedlings, longer exposure time to adverse conditions, such as the action of birds, insects, diseases, in addition to greater risk of loss of seeds by deterioration and thus reduce productivity and, consequently, producer profit (Eira et al. 1993).

The knowledge of dormancy mechanisms and their duration for different vegetable seeds are of ecological and economic importance, since they help in choosing the specific treatments to act on the seed metabolism, releasing the embryo for development or making it able to germinate (Lopes 2012).

In the germination process of a seed, it is necessary that physical and chemical environments are favorable, in other words, that there is availability of water, temperature and concentration of oxygen in the medium that do not limit the germination metabolism of the seeds. However, some vegetable seeds such as lettuce, carrot, eggplant, okra, beet, pea do not germinate even in favorable environmental conditions. These seeds are called dormant, because they present some internal or systemic restriction that prevents the embryo development (Lopes 2012).

The dormancy phenomenon allows distribution of germination of the seeds in time, and is also an adaptive characteristic that ensures perpetuation and survival of the species in different ecosystems. Germination retardation maintains the species at seed stage, that is, at the stage of the plant cycle where its resistance to unfavorable conditions is greater (Vivian et al. 2008).

Seed dormancy is controlled by the genotype associated with environmental factors, such as temperature; physical factors, such as tegument thickness; and regulation of hormones, such as abscisic acid and gibberellin. Generally, plants produce seeds with various levels of dormancy. This aspect helps to distribute their descendants through time and space. In addition, it allows the seeds not to germinate during the adverse periods, guaranteeing a time required for the seed dispersion by a greater geographic distance providing the species survival (Poisot et al. 2011). The type of seeds dormancy, therefore, can influence processes of colonization, adaptation, specialization and extinction of the species (Willis et al. 2014).

The seed position in the plant also influences the occurrence of dormancy. This aspect is observed in carrot, where the seeds of the primary umbels are heavier and with deeper dormancy (Lopes 2012). Some factors may interfere with germination of seed physiological quality. If the maturity of the seeds occurs before the fruit maturity, and thus losing its quality when the fruits are completely mature and dehiscent, fruit harvest should be earlier aiming to use the seed for the production of seedlings (Kosera Neto et al. 2015).

Aiming to standardize and increase the germination of some seeds it is necessary to identify the type of dormancy, and then to define methods for breaking dormancy, which aims to decrease the tegument and allow the absorption of water (Lorenzi 2002).

2. Dormancy Categories

The dormancy is usually classified according to its origin or with the probable mechanisms involved. Regarding the origin, two distinct categories are currently recognized: primary or natural and secondary or induced dormancy.

2.1. Primary or Natural

Primary dormancy sets in the stage of development and/or maturation of the seeds, while they are still physiologically bound to the mother plant, preventing germination of the seed after harvest. The stage of seed development and maturation in which dormancy is established varies among species.

The mechanisms that lead to the absence of germination occur due to several factors, such as physiological, physical, morphological, chemical, genetic, mechanical immaturity or the combination of these.

This type of dormancy is naturally overcome after a few days, or in a few months after harvest, and there is no need for treatments (Kano et al. 2011). Eggplant seeds, for example, do not germinate soon after harvest, and the phenomenon was overcome during storage of the dried seed for a period not too long (Freitas and Nascimento 2006).

The natural dormancy is important for the species perpetuation, allowing the survival of seeds in unfavorable conditions, and prevent germination in the mother plant, a phenomenon known as viviparity.

2.2. Secondary or Induced

Secondary dormancy occurs when the seed is already apt for germination. After reaching maturity, soon after the harvest or dispersion, germination blockage can occur when undergoing a period of environmental stress, mainly related to water, temperature, light and oxygen factors, which induce a mechanism of protection in the seed, preventing the germination in unfavorable conditions for their survival.

This process involves several steps and physiological, physical and/or chemical changes in the seed and can be established in two stages: in the imbibition - phase that precedes the germination of the seed, and in storage, when it is induced with prolonged periods of high temperatures.

- Thermo-inhibition: a reversible process in which the temporary inability of seed germination is due to unfavorable temperatures. It is overcome in a simple and natural way by establishing suitable temperatures for germination, with no specific treatments.
- Thermo-dormancy: is considered a profound form of inactivity due to exposure to prolonged periods at high temperature. It needs treatment to get out of this condition, and the seeds do not germinate after the temperature is reduced. Moisturizing and dehydration treatments, known as osmotic conditioning "priming" of seeds prone to thermo-dormancy can avoid problems with germination.

Non dormant seeds of lettuce, for example, may acquire secondary dormancy when placed to germinate at high temperatures. Primary dormancy and secondary dormancy are independent processes, since the seed may acquire secondary dormancy without being dormant or after having overcome primary dormancy (Lopes 2012).

3. Causes of Dormancy

Seed dormancy does not occur due to a single cause or mechanism, although the final consequence of the process is the same. It is essential to know the cause of dormancy so that one can plan or choose a method to overcome it, or understand why the seeds of various species of vegetables do not germinate uniformly.

Seed dormancy is classified into two major groups: endogenous dormancy caused by some blockage to germination imposed by the tissues surrounding the embryo; and the exogenous one that is the dormancy caused by the seed coatings (tegument, endocarp and pericarp) and is related to impermeability, the mechanical effect and/or the presence of substances inhibiting the tissues (Cardoso 2004). In both situations, the embryo is unable to overcome the restrictions imposed on it, in the first by closure of the tissues and in the second by the internal factors of the embryo (Lopes 2012).

4. Some of the Main Types of Dormancy

4.1. Immature or Rudimentary Embryo

Seed embryos of several species are morphologically immature at the time of plant seed dispersal and require a certain period of time for their complete development. Only after this interval, the seeds will be able to trigger the germinative process. Immature embryos, in general, are relatively underdeveloped and, in some cases, not totally differentiated. There is usually a need for a period of three to four months to complete its development.

4.2. Waterproof Tegument

Seeds with waterproof tegument are called hard seeds. From the point of species perpetuation, this characteristic is extremely beneficial for keeping the seeds alive for a long period. It is the most common cause of dormancy in plant families found in arid and semi-arid regions, as the seed tegument (cover) prevents water absorption. The rupture of this is immediately followed by imbibition and beginning of the germinative process. They occur in the families Fabaceae (pod-beans and peas), Malvaceae (okra) and Convolvulaceae (sweet potatoes) (Lopes 2012). The rate and degree of drying of the seeds are important in this mechanism, because the teguments become progressively hard and impermeable as the degree of humidity decreases.

4.3. Impermeability to Gas Exchanges

The teguments of some seeds present restrictions imposed by the tegument to oxygen absorption, the release of carbon dioxide or both mechanisms. This low permeability suggests that the seed tegument can inhibit germination by limiting the supply of this gas for the embryo respiration.

This mechanism occurs in seeds of beet (ovary wall), lettuce (pericarp or endosperm wall), pumpkin and celery (Lopes 2012). By means of scarifying, cutting, treatment with acids or subjecting the seeds to conditions of high oxygen tension to increase the percentage of germination of these seeds.

4.4. Mechanical Restrictions

The seed tegument and other enveloping structures of the embryo are extremely important, both in survival and regulation of the seed germination process. Because the seed tegument is the embryo's primary defense against dangers posed by environmental variations, it is not surprising that it can impose barriers to rapid and early germination. Therefore, the protective structures that surround the embryo can present two functions, that is, protection of the embryonic axis and regulation of the time of seeds germination.

In some cases, the seed tegument may be so rigid that it does not allow the passage of the radicle. Solid and lignified shells are factors responsible for mechanical hardness. Such tissues must be broken down by biotic or environmental forces so that the seed can germinate.

Many non-lignified tissues such as the lettuce seed endosperm can suppress the expansion of the embryo. In this case, for germination to occur, the endosperm cell walls must be broken down by cell wall degrading enzymes. It has been clearly demonstrated in lettuce (Nascimento et al. 2004) and tomato (Still and Bradford 1997) where the conditions for breaking dormancy are related to the weakening of endosperm cell walls near the radicle. This weakening occurs through the action of the enzyme endo-b-mannanase; this is because cell wall cells are rich in manose.

4.5. Dormant Embryo

Characterized by the fact that the embryo itself is the seat of dormancy, it is the result of physiological conditions in the embryo and presents special requirements as to light and cooling to overcome dormancy, or whose causes are chemical inhibitors. These substances act in interaction with environmental factors, such as temperature and oxygen availability.

4.6. Physical Dormancy

It is characterized by the presence of extra-embryonic tissues that prevent the entry of water or gas exchange between the embryo and the environment. This is the type of dormancy imposed on the embryo by the seed tegument or by other tissues that surround it, such as endosperm, pericarp or extra-floral organs (Baskin and Baskin 2004).

Hydration and consequent interruption of dormancy are in many cases related to the formation of openings in specialized anatomical structures (eg, hilum and lens) located on the surface of the seed, causing a decrease in resistance to water ingress (Baskin and Baskin 2004).

It is important to determine the exact function of teguments and/or wraps in the diaspore germination, before attempting to classify the seed accordind to the dormancy nature, which may be exogenous (imposed by the wraps), and associated with the following causes:

4.7. Physiological Dormancy (PD)

PD is basically regulated at metabolic and gene levels. This class is divided into three levels - deep, intermediate and not deep - being the deep PD characterized by the inability of the isolated embryo to produce a normal seedling, whereas at the intermediate and not deep levels, the excision of the embryo is sufficient to make it develop and produce a normal seedling. Therefore, in PD operates different mechanisms associated with the embryo, as well as adjacent tissues and structures, such as the tegument and endosperm.

The differences between the intermediate and non-deep levels refer basically to the treatments required to break dormancy. The intermediate PD, for example, requires a cold stratification period of 2 to 3 months, and may reach 6 months in some species; while in the non-deep PD, relatively short periods of stratification (ranging from 5 to 90 days depending on the species) may be sufficient to break dormancy. In addition, in some species the non-deep PD can be disrupted with exposure to higher temperatures (≥15°C) than those normally employed in cold stratification (4 - 6°C) (Baskin and Baskin 1997).

According to Baskin and Baskin (2004), in physiological dormancy, the presence of inhibitory substances or absence of promoting substances prevents germination from occurring, meaning, it is inherent to the embryo and is not due to any influence of the tegument or of another seed tissue.

The susceptibility to chemical regulators and the phytochrome photo-equilibria are some of the factors that must be located only in the embryo, and factors related to the extra-embryonic tissues also must be considered (Cardoso 2004).

Giberelic acid (GA) is an essential phytohormone that controls many aspects of plant development, including seed germination, leaf expansion, stem elongation, flowering and seed development, being used to overcome physiological dormancy (Davies 2004).

4.8. Morphological Dormancy (MD)

It is manifested in seeds that are released from the mother plant with differentiated embryos (cotyledons and recognizable hypocotyl-radicle axis), but underdeveloped in size. In this case, "visible" (protrusion) germination is preceded by an intra-seminal growth phase triggered by appropriate environmental conditions (Cardoso 2009).

According to Baskin and Baskin (2004) such seeds do not respond to dormancy treatments, following only their own growth rate. Interestingly, when accompanied by physiological dormancy, MD appears to be sensitive to dormancy-breaking treatments. The cases of seeds dispersed with the undifferentiated embryo were no longer classified as MD due to the fact that their development (histo-differentiation phase) was not completed and therefore they were considered immature seeds.

4.9. Morphophysiological Dormancy (MPD)

In MPD besides the underdeveloped embryo, there is a physiological component that requires treatments or conditions to break dormancy. Depending on the species, the growth is either preceded by the PD break, or both processes (PD break and growth) occur simultaneously (Baskin and Baskin, 1998). The differences between the various levels and types of MPD are related to the methods used for dormancy breaking, based either on thermal treatments (stratification) or treatments with gibberellic acid.

5. Dormancy Release Methods

The method to be used to overcome the different causes depends on the type of dormancy present in the seeds. Some methods for overcoming dormancy are imbibition, mechanical scarification, acid scarification, thermal scarification, washing in running water, pre-drying, precooling, stratification, chemicals, alternating temperatures and exposure to light to weaken its tegument and allow the absorption of water (Lorenzi 2002). According to Martins and Silva (2003), a promising possibility that has been studied in research aimed to overcome dormancy, is the use of thermal treatments.

5.1. Hydroconditioning

Satisfactory results on the percentage of germination after immersion of seeds in water at room temperature, can be explained by the increase in the size of its size due to imbibition. The hydrated tegument softens and breaks and; the growth tissues develop with the supply of food by the cotyledons (Souza et al. 2012).

Work done by Silva et al. (2005) found a result of 92% germination, when the beet seeds were immersed for two hours. Evaluating the germination of Cherry tomato seeds (S. pseudocapsicum L.), using thermal treatments to overcome dormancy, De Nardin et al. (2014) found that the best method was immersion in water at 25°C for a period of 12 hours, with 42% germination.

Nascimento (2014) reports that a treatment with pre-soaking in water, for a period of 24 to 48 hours, can minimize and improve the germination of okra seeds. With the study of different imbibition times and concentrations of gibberellin in sweet passionflower seeds (*Passiflora alata Curtis*), Ferreira et al. (2001) observed 71.2% germination at the concentration of 500 mg L^{-1} of GA_3, whereas the control presented only 56%.

Pereira et al. (2008) working with carrot seeds (*Daucus carota* L.) cv. Brasilia verified that the osmotic conditioning in PEG 6000 for four days, independent of the osmotic potential and the imbibition method, contributed to increase the germination of the seeds of the two lots by up to 20%, reaching 94% (lot 1) and 92% (lot 2).

5.2. Acid Scarification

This method consists in submerging the seeds in concentrated sulfuric acid for a certain period of time and then washing under running water and drying. The submergence time in the acid is critical and must be carefully determined, and may vary from a few minutes to several hours, depending on the species. This type of scarification is recommended to overcome dormancy of seeds that have teguments impermeable to water or gases, such as strawberry achenes (Galvão et al. 2014).

Stenzel et al. (2003) in work carried out on *Annona squamosa* L. the seeds soaked with gibberellic acid (GA_3) at 50 ppm provided the highest percentage of germination, compared to GA_3 treatment at 100 ppm in water (5°C and 30°C) and control.

5.3. Thermal Scarification

The treatment consists in immersing the seeds in water with temperature ranging from 60 to 100°C, during a predetermined time interval, which is variable according to the species to be treated.

Treatment with water at 60°C and osmoconditioning with PEG 6000 (200 g L^{-1} Polyethylene glycol) increased germination rate and uniformity, as well as seedling growth, both in the laboratory and in the field, in the seeds of *Adesmia latifolia* conditioned for two days (Sune et al. 2002). In work with chili seeds (*Capsicum annuum* L.), Posse et al. (2002) found improvement in the germination process with water conditioning being more effective than osmoconditioning.

Hölbig et al. (2011), when evaluating the physiological performance of onion seeds of cv. Bola Precoce, verified that hydroponic and fungicide treatment increased the emergence rare and the percentage of emerged seedlings.

5.4. Washing in Current Water

This method is used for species whose seeds have some water-soluble inhibitory substances, which can be removed by simply washing the seeds in running water for a variable time interval according to the species. Silva et al. (2005) found a significant effect on the beet seed by the immersion period of two hours, with a germination percentage of 92%.

5.5. Pre-Cooling

Seeds of some species of Brassicas overcome dormancy when submitted to a temperature of 5 to 10°C for a variable time according to species. The technique of pre-planting vernalization, developed by Brazilian researchers, consists in subjecting the garlic seed in a cold chamber with a temperature of 3 to 5°C, for a period of 40 to 60 days allowing the cultivation of noble garlic cultivars in Brazillian Southern, in regions where thermo-photoperiodic conditions do not meet the requirements of the plant (Souza and Macedo, 2009). Resende et al. (2011), evaluating two garlic cultivars submitted to three vernalization periods (30, 40 and 50 days), verified that the 40-day period allowed higher growth and satisfactory productivity in the region of Guarapuava, Paraná State, Brazil.

5.6. Mechanical Scarification

Among the methods used to overcome the tegumentary dormancy, mechanical scarification is a frequently used technique and is the most practical and safe option for small farmers (Hermansen et al. 2000), as well as being a simple, low-cost and effective method to promote rapid and uniform germination.

It consists of subjecting the seeds against abrasive surfaces, such as sandpaper or silicon carbonate stone, with the aim of degassing or eliminating part of the seeds tegument. Scarification should not be too severe, as it may cause injury to the embryo, impairing germinative performance. Mechanical scarification is basically used to overcome dormancy of seeds that have a water-impermeable tegument (hard seeds).

Based on the results found by Santos et al. (2004), to overcome dormancy of *Sterculia striata* (Monkey's peanut) seeds, it is recommended scarification with sandpaper n°40 on one side of the seed for 3 minutes, followed by imbibition in water for 24 hours, and mechanical scarification with sandpaper n° 40 on both sides of the seed, for 3 minutes, without imbibition, with 70% and 60% germination, respectively.

5.7. Exposure to Light

Seeds of many species are light sensitive (photoblastic positive) and do not germinate in their absence or exhibit very reduced germination in this condition. Lettuce seeds is an example of a species whose seeds are sensitive to light (Andrade 1995).

The seeds can be classified, according to the presence of light, in non-photoblastic (not influenced by light), relative positive photoblastic (when the percentages of germination are significantly higher in the light, but not more than double of the dark regime), positive photoblastic (when the percentage of germination in light is greater than twice the percentage in the dark) or even relative negative photoblastic (when they germinate best in the dark) (Ferreira, et al 2001).

5.8. Stratification

One of the methods used to promote the overcoming of physiological dormancy is the use of stratification (Debska et al. 2013), which consists of keeping seeds at low temperatures (Keshavarzian et al. 2013), or the exogenous application of solutions containing gibberellins.

Giberelic acid, considered an endogenous enzymatic activator, can promote seed germination when exogenously applied (Dalastra et al. 2010). Under natural conditions, apple seed dormancy is removed by 90-day stratification (5°C) (Lewak 2011). This treatment leads to changes in GAs, ABA, ethylene and PA content (Lewak 2011), which are accompanied by fluctuation in H_2O_2 concentration in the tissue and production of NO (Dębska et al. 2013).

Kosera Neto et al. (2015) reported that *Solanum betaceum* seeds showed higher percentage of germination when exposed to light, regardless of the method of overcoming dormancy. However, the use of gibberellic acid at the concentration of 300 mg L^{-1} resulted in a superior germination rate.

5.9. Temperature Alternation

The alternation of temperature, which occurs naturally under field conditions, seems to act on seed teguments making them more permeable to water and oxygen and may also have an influence on the balance between the promoting and inhibiting substances of germination.

In a trial with *Cumari Capsicum baccatum* var. *praetermissum*, seeds were immersed in hot water (38°C) for 10 seconds and then dipped in cold water for another 10 seconds. The germination of the seeds was superior in the treatment with alternating thermal shock in hot water and cold water with 100% germination up to 60 days.

5.10. Post-Maturation

It occurs when the seeds do not germinate due to the immaturity of the embryo. In this case, an additional period is necessary for its complete development, called post-maturation (Ferreira and Borghetti 2004). According to Cardoso (2004) some seeds present post-maturation after passing through a period of low temperatures, while others require temperatures in the range of 35 to 40°C dry and there are still non-hydrated seeds kept at elevated temperatures (40 to 60°C) for a few days or several months.

Conclusion

Based on the studies carried out by several authors, it can be seen that there are still aspects to be elucidated regarding the best and most efficient methods for breaking dormancy in vegetable seeds. Further studies should be carried out considering several aspects that may interfere, such as the species, the origin of the material, the harvesting time, the storage time, and the physicochemical characteristics of the propagules and seeds.

REFERENCES

ABCSEM. (2016). *Associação Brasileira do Comércio de Sementes e Mudas* [*Brazilian Association of Seed and Seed Trade*]. http://www.abcsem.com.br/.

Andrade, A. C. S. (1995). Efeito da luz e da temperatura na germinação de *Leandra breviflora Cogn., Tibouchina benthamiana Cogn., T. grandiflora Cogn. e T. moricandiana (DC.) Baill.* (Melastomataceae). [Effect of light and temperature on the germination of *Leandra breviflora Cogn., Tibouchina benthamiana Cogn., T. grandiflora Cogn. and T. moricandiana (DC.) Baill.* (Melastomataceae)] *Revista Brasileira de Sementes,* 17: 29-35.

Baskin, C. C. and Baskin, J. M. (1998). *Seeds: ecology, biogeography and evolution of dormancy and germination.* Academic Press, San Diego. 666p.

Baskin, J. M. and Baskin, C. C. (1997). Methods of breaking seed dormancy in the endangered species Iliamna corei (Sherff) Sherff (Malvaceae), with special attention to heating. *Natural Areas Journal,* 17: 313–323.

Baskin, J. M. and Baskin, C. C. (2004). Classification, biogeography, and phylogenetic relationships of seed dormancy. In Pritchard, H. (Ed.) *Seed conservation: turning science into practice.* Kew, The Royal Botanic Gardens (in press).

Camargo Filho, W. P. and Camargo, F. P. (2017). A produção e a comercialização mundial dos principais olerícolas [The production and the worldwide commercialization of the main olerícolas]. *Horticultura Brasileira,* Brasília, 35: 1-2.

Cardoso, V. J. M. (2004). Dormência: estabelecimento do processo [Dormancy: establishment of the process]. In: A. G. Ferreira and F. Borghetti (orgs). *Germinação: do básico ao aplicado.* Artmed, Porto Alegre. 323p.

Cardoso, V. J. M. (2009). Conceito e classificação da dormência em sementes [Concept and classification of seed dormancy]. *Oecologia Brasiliensis* 13: 619-631.

Dalastra, I. M., Pio, R., Entelmann, F. A., Werle, T., Uliana, M. B. and Scarpare Filho, J. A. (2010). Germinação de sementes de nogueira-macadâmia submetidas à incisão e imersão em ácido giberélico [Germination of walnut-macadamia seeds submitted to incision and immersion in gibberellic acid]. *Ciência e Agrotecnologia,* 34: 641-645.

Davies, P. J. (2004). *Plant Hormones: Biosyntesis, Signal Transduction, Action!,* Kluwer Academic, Netherlands.

Debska, K., Krasuska, U., Budnicka, K., Bogatek, R. and Gniazdowska, A. (2013). Dormancy removal of apple seeds by cold stratification is associated with fluctuation in H_2O_2, NO production and protein carbonylation level. *Journal of Plant Physiology,* 170: 480-488.

De Nardin, G. G., Kutz, T. S., Cardoso, J., Pasini, R., Vargas, T. O, Brum, B. and Godoy, W. I. (2014) *Congresso Paranaense de Agroecologia – Cadernos de Agroecologia* [*Paranaense Congress of Agroecology – Agroecology Journals*], 9, Pinhais-PR.

Eira, M. T. S., Freitas, R. W. A. and Mello, C. M. C. (1993). Superação da dormência de sementes de E*nterolobium contortisiliquum* (Vell.) Morong. – Leguminosae [Overcoming seed dormancy of Enterolobium contortisiliquum (Vell.) Morong. – Leguminosae]. *Revista Brasileira de Sementes,* 15: 177-181.

Ferreira, G., Fogaça, L. A. and Moro, E. (2001). Germinação de sementes de *Passiflora alata Dryander* (maracujá-doce) submetidas a diferentes tempos de embebição e concentrações de ácido giberélico [Germination of seeds of Passiflora alata Dryander (sweet passion fruit) submitted to different soaking times and concentrations of gibberellic acid]. *Revista Brasileira de Fruticultura*, 23:160-163.

Ferreira, A. G. and Borghetti, F. (2004). *Germinação: do básico ao aplicado* [*Germination: from basic to applied*]. Porto Alegre: Artmed, 323 p.

Freitas, R. A. and Nascimento, W. M. (2006). Teste de envelhecimento acelerado em sementes de lentilha [Test of accelerated aging in lentil seeds]. *Revista Brasileira de Sementes,* 28: 59-63.

Galvão, A. G., Resende, L.V., Guimaraes, R. M., Ferraz, A. K. L., Morales, R. G. F., Marodin, J. C. and Catão, H. C. R. M. (2014). Overcoming strawberry achene dormancy for improved seedling production in breeding programs. *Idesia*, 32: 57-62.

Hermansen, L. A., Duryea, M. L., West, S. H., White, T. L. and Malavasi, M. M. (2000). Pretreatments to overcome seed coat dormancy in *Dimorphandra mollis*. *Seed Science & Technology*, 28: 581-595.

Hölbig, L. S., Baudet, L. and Villela, F. A. (2011). Hidrocondicionamento de sementes de cebola [Hydrodeposition of onion seeds]. *Revista Brasileira de Sementes*, 33: 171 - 176.

Kano, C., Cardoso, A. I. I., Bôas, R. L. V. and Higutti, A. R. O. (2011). Germinação de sementes de alface obtidas de plantas cultivadas com diferentes doses de fósforo [Germination of lettuce seeds obtained from plants grown with different doses of phosphorus]. *Semina: Ciências Agrárias*, 32: 591-598.

Keshavarzian, M., Gerivani, Z., Sadeghipour, H. R., Aghdasi, M. and Azimmohseni, M. (2013). Suppression of mitochondrial dehydrogenases accompanying post-glyoxylate cycle activation of gluconeogenesis and reduced lipid peroxidation events during dormancy breakage of walnut kernels by moist chilling. *Scientia Horticulturae*, 161: 314-323.

Kosera Neto, C., Fabiane, K. C., Radaelli, J. C., Wagner Júnior, A. and Moura, G. C. (2015). Métodos para superação de dormência em sementes de tomateiro arbóreo (*Solanum betaceum*). [Methods for overcoming dormancy in arboreal tomato seeds (Solanum betaceum)]. *Pesquisa Agropecuária Tropical*, 45: 420-425.

Lewak, S. (2011). Metabolic control of embryonic dormancy in apple seed: seven decades of research. *Acta Physiologiae Plantarum*, 33:1-24.

Lopes, A. C. A. (2012). *Dormência em sementes de hortaliças* [*Dormancy in vegetable seeds*]. Brasília, DF: Embrapa. ISSN 1415-2312.28 p. – (Documentos / Embrapa Hortaliças; 136).

Lorenzi, H. (2002). *Árvores Brasileiras: manual de identificação e cultivo de plantas arbóreas do Brasil* [*Brazilian Trees: a manual for identification and cultivation of tree plants in Brazil*]. 2. ed. Nova Odessa: Instituto Plantarum, 368 p.

Martins, L. and Silva, W. R. (2003). Efeitos imediatos e latentes de tratamentos térmico e químico em sementes de *Brachiaria brizantha* cultivar Marandu [Immediate and latent effects of thermal and chemical treatments on seeds of *Brachiaria brizantha* cultivar Marandu]. *Bragantia*, 62: 81-88.

Nascimento, W. M., Cantliffe, D. J. and Huber, D. J. (2004). Ethylene evolution and endo-βmannanase activity during lettuce seed germination at high temperature. *Scientia Agricola,* 61: 156-163.

Nascimento, W. M. (2014). *Produção de sementes de hortaliças* [*Production of vegetable seeds*]. Brasília, DF: Embrapa.

Pereira, M. D., Dias, D. C. F., Dias, L. A. S. and Araújo, E. F. (2008). Germinação e vigor de sementes de cenoura osmocondicionadas em papel umedecido e solução aerada [Germination and vigor of osmoconditioned carrot seeds in moistened paper and aerated solution]. *Revista Brasileira de Sementes*, 30: 137-145.

Poisot, T., Bever, J. D., Nemri, A., Thrall, P. H. and Hochberg, M. E. (2011). A conceptual framework for the evolution of ecological specialisation. *Ecology Letters,*14: 841-851.

Posse, S. C. P., Silva, R. F., Vieira, H. D. and Catunda, P. H. A. (2002). Efeitos do condicionamento osmótico e da hidratação na geminação de sementes de pimentão (*Capsicum annuum* L.) submetidas à baixas temperaturas [Effects of osmotic conditioning and hydration on the twinning of chili seeds (*Capsicum annuum* L.) subjected to low temperatures]. *Revista Brasileira de Sementes,* 23: 123-127.

Resende, J. T. V., Morales, R. G. F., Resende, F. V., Faria, M. V., Souza, R. J. and Marchese, A. (2011). Garlic vernalization and planting dates in Guarapuava. *Horticultura Brasileira,* 29: 193-198.

Santos, T. O., Morais, T. G. O. and Matos, V. P. (2004). Escarificação mecânica em sementes de chichá (*Sterculia foetida* L.) [Mechanical scarification in chichá seeds (*Sterculia foetida* L.)]. *Revista Árvore,* 28: 1-6.

Scalon, S. P. Q., Mussury, R. M., Watheir, F., Gomes, A. A., Silva, K. A., Pierezan, L. and Scalon Filho, H. (2005). Armazenamento, germinação de sementes e crescimento inicial de mudas de *Enterolobium contortisiliquum* (Vell.) Morong [Storage, seed germination and initial growth of *Enterolobium contortisiliquum* (Vell.) Morong seedlings]. *Acta Scientiarum. Biological Sciences*, 27: 107-112.

Silva, J. B., Vieira, R. D. and Cecílio Filho, A. B. (2005). Superação de dormência em sementes de beterraba por meio de imersão em água corrente [Overcoming of dormancy in beet seeds by means of immersion in running water]. *Horticultura Brasileira,* 23: 990-992.

Souza, R. J. and Macêdo, F. S. (2009). Botânica e Cultivares [Botany and Cultivars]. In. *Cultura do alho: Tecnologias Modernas de Produção.* Lavras: UFLA, p. 21-37.

Souza, T. V., Voltolini, C. H., Santos, M. and Paulilo, S. M. T. S. (2012). Water absorption and dormancy breaking requirements of physically dormant seeds of *Schizolobium parahyba* (Fabaceae – Caesalpinioideae). *Seed Science Research,* 22: 169-176.

Stenzel, N. M. C., Murata, I. M. and Neves, C. S. V. J. (2003). Superação da dormência em sementes de atemóia e fruta-do-conde [Overcoming dormancy in seeds of atemóia and fruit-of-earl]. *Revista Brasileira Fruticultura,* 25: 305-308.

Still, D. W. and Bradford, K. J. (1997). Endo-beta-mannanase activity from individual tomato endosperm caps and radicle tips in relation to germination rates. *Plant Physiology,* 113: 21-29.

Sune, A. D., Franke, L. B. and Sampaio, T. G. (2002). Efeitos do condicionamento osmótico na qualidade fisiológica de sementes de *Adesmia latifólia* (Spreng.) [Effects of osmotic conditioning on the physiological quality of Adesmia latifolia (Spreng.) seeds]. *Vog. Revista Brasileira de Sementes,* 24: 18-23.

Tedesco, S. B., Stefanello, M. O., Schifinowittmann, M. T., Battistin, A. and Dall'agnol, M. (2001). Superação da dormência em sementes de espécies de Adesmia D.C. (Leguminosae) [Overcoming dormancy in seeds of Adesmia D.C. (Leguminosae) species]. *Revista Brasileira de Agrociência,* 7: 89-92.

Vivian, R., Silva, A. A., Gimenes Júnior, M., Fagan, E. B., Ruiz, S. T. and Labonia, V. (2008). Dormência em sementes de plantas daninhas como mecanismo de sobrevivência [Dormancy in weed seeds as a survival mechanism]. *Planta Daninha,* 26: 695-706.

Willis, C. G., Baskin, C. C., Baskin, J. M., Auld, J. R., Venable, D. L., Cavender-Bares, J., Donohue, K. and Casas, R. R. (2014). The evolution of seed dormancy: environmental cues, evolutionary hubs, and diversification of the seed plants. *New Phytologist,* 203: 300-309.

In: Plant Dormancy
Editor: Renato V. Botelho

ISBN: 978-1-53615-380-4

Chapter 8

OVERCOMING SEED DORMANCY OF FOREST SPECIES IN MIXED OMBROPHYLOUS FORESTS

Luciano Farinha Watzlawick**, *Joelmir Augustinho Mazon* and *Richeliel Albert Rodrigues Silva
[1]Department of Agronomy Midwestern Parana State University
Guarapuava, PR, Brazil

ABSTRACT

The Mixed Ombrophilous Forest (MOF), also known as Araucaria Forest, is a phytophysiognomy located within the Atlantic Forest, the fourth most important biodiversity hotspot on the planet. Due to the great diversity of species present in this phytophysiognomy, it is important to know its ecological characteristics, as is the case of seed dormancy. In the MOF, most of the seeds are classified as orthodox, present physical dormancy, low to moderate germination. Several techniques are used to overcome dormancy of forest seeds. The procedure is performed to

* Corresponding Author's E-mail: farinha@unicentro.br.

accelerate germination process, among techniques, chemical and mechanical scarification are the most applied in seeds of the MOF species, due to high percentage of seeds with physical dormancy. It is verified that the majority of species that present dormancy belongs to the families Fabaceae and Lauraceae, mainly the species of the genus *Ocotea*, that are present in the forest upper stratum. In addition, some seeds require pre-germination treatment, even if they do not present dormancy. Therefore, the information on seed dormancy of the species of the MOF generating an important tool for better decision making in management of forest recovery techniques.

1. The Mixed Ombrophylous Forest

Many neotropical regions harbor some of the planet's most diverse ecosystems, but many of these ecosystems are unknown to researchers and knowledge about how much and why these ecosystems are so diverse is still scarce.

The Mixed Ombrophilous Forest, also known as Araucaria Forest, is a phytophysiognomy located within the Brazilian Biome "Atlantic Forest," the fourth most important biodiversity hotspot on the planet. This subtropical forest ecosystem is highly threatened due to human and agricultural demographic expansion in southern Brazil that has been occurring over the past two centuries.

This vegetation occurred more frequently in the Southern Highlands of Southern Brazil (Figure 1) and its original composition is a result of the meeting of Afro-Brazilian (tropical) and Austral-Antarctic-Andean (temperate) flora, with emphasis on Coniferales and Laurales and genera *Araucaria, Podocarpus, Drymys, Ocotea* and *Nectandra*, connoting the term of "mixed."

The Mixed Ombrophilous Forest has its occurrence delimited by the region of temperate climate of the South of Brazil, in altitudes between 400 m and 1,200 m s.m.m. In this region the pluviometric regime is high and well distributed throughout the year. Annual records of less than 1,400 mm of rainfall are rare, exceeding 2,000 mm in some years or at some points in rivers and mountains, justifying the term "ombrophilous," from the Greek "rain-loving." The forest coincides with the isothermal region of 18 °C, practically the same isotherm region of southern Brazil, presenting frequent frosts, with annual average lower than 18 °C, with a minimum lower than 10 °C in the winter months and higher than 20 °C in the summer months. The occurrence of this phytophysiognomy is probably limited by the seasonality of regular rainfall and low temperatures.

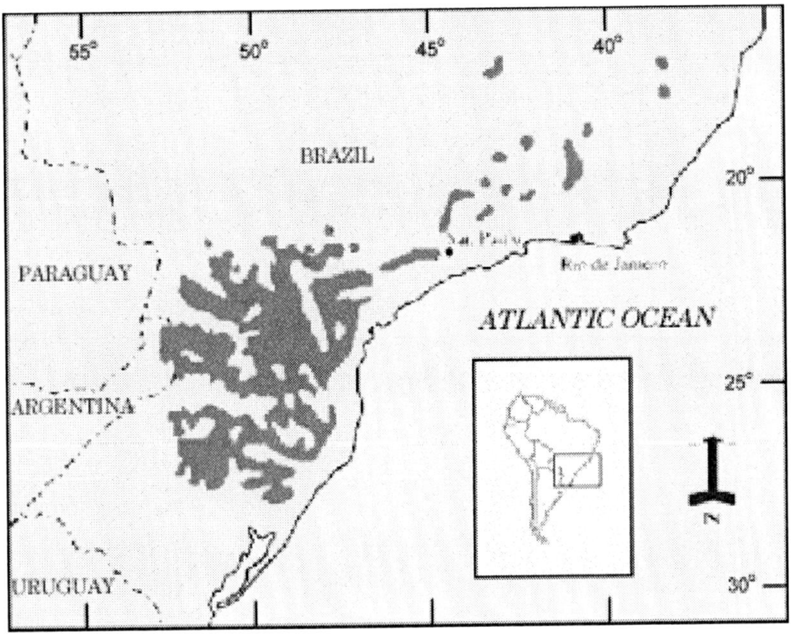

Source: Conifer Society, 2018.

Figure 1. Original occurrence of Mixed Ombrophilous Forest in Brazil.

The Mixed Ombrophilous Forest would have expanded over the Holocene period (11,500 years ago to present times) more timidly over the fields, confined to the river valleys, which was probably the result of a dry climate and higher temperatures, occurring only during the middle and upper Holocene times the increase of the representation of genera *Araucaria, Dicksonia, Podocarpus*, among others in the phytophysiognomy. This fact suggests, given the altitude and latitude of the Southern Highlands, a recent occupation from Alto-Montane Refuges, finding its current climatic climax.

In addition to the climatic system, evidence has been obtained that pre-cabralian indigenous populations may have contributed directly or indirectly to the implantation of areas with the iconic-species *Araucaria angustifolia* (Bertol.) Kuntze populations, precisely at a time when the climate was not favorable. This fact is evidenced because that some forests appear to have expanded precisely in times when there was an increase in the complexity and size of populations of ancestral indigenous groups called *Proto-Jê*, about 4,000 to 3,000 years ago.

These morpho-climatic aspects shaped the forest into four distinct formations: Subtropical Ombrophilous Alluvial Forest, on old terraces associated with hydrographic network; the Lower Montane Forest, constituting disjunctions at altitudes below 400 m; the Montane Forest, located approximately between 400 and 1,000 m of altitude; and Upper Montane Forest at altitudes above 1,000 m.

Araucaria angustifolia (Bert.) O. Kuntze (Parana-pine) is the only representative of the family Araucariaceae in Brazil and, along with related species *Araucaria araucana* (Molina) K. Koch (Monkeypuzzle tree), are the only existing representatives of the family in American continent. *Araucaria angustifolia* plays a prominent role in Mixed Ombrophilous Forest, because it is the iconic species and determinant for its phytophysiognomy because it occupies the upper stratum of forest and has high density. However, due to its high economic value, this species has been intensively explored over the two last centuries and it is included in vulnerable category in the list of endangered species of Brazil.

Araucaria angustifolia populations create a microecosystem from the crown dominance of their adult individuals providing shade to shade-tolerant species, guaranteeing adequate conditions for these and other species to develop. Their seeds, which are popularly called “pinhões”, also feed the wild fauna during the period of greatest food shortage in the forest, which is from late February to July, and can last for a few more months. These seeds have great nutritional value as part of the wild fauna, from birds to mammals that are their main dispersers, besides being highly appreciated and result in an alternative income for the region people.

The Mixed Ombrophilous Forest presents a well defined and stratified structure, with an emerging stratum naturally occupied and practically exclusively by the *Araucaria angustifolia*, reaching heights greater than 25 meters and presenting individuals up to 40 meters. The canopy generally reaches 20 meters in height (Figure 2), being predominantly occupied by woody species of Myrtaceae and Lauraceae, in addition to common presence of *Podocarpus lambertii* Klotzsch ex Endl, one of the few gymnosperms in this forest.

The Mixed Ombrophilous Forest presents about 350 tree species, with a high degree of endemism. The Myrtaceae family presents greater representativeness of species, followed mostly by Lauraceae, with the occurrence of Sapindaceae and Aquifoliaceae in more conserved fragments, whereas the species of Asteraceae, Euphorbiaceae and Fabaceae are more common in riparian or secondary forests.

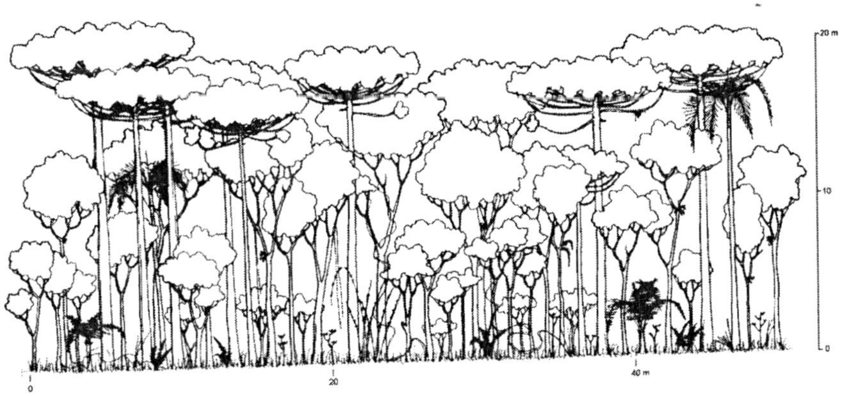

Figure 2. A schematic profile highlighting the structure of a segment of Mixed Ombrophilous Forest, with predominance of genera such as *Araucaria, Ocotea, Cedrela, Casearia, Sloanea, Podocarpus, Campomanesia, Ilex* and *Cinnamodendron* (Adapted from Roderjan et al. 2002).

The most representative and important species that occur in a high degree of association with *Araucaria angustifolia* in this phytogeographic unit are: *Ocotea porosa* (Nees & Mart.) Barroso (Brazilian-walnut) and *Ilex paraguariensis* A. St.-Hil. (Yerba-mate), besides Lauraceae like those of the genus *Nectandra* and *Ocotea*, the Fabaceae of the genus *Dalbergia* and

Machaerium, the Meliaceae mainly *Cedrela fissilis* Vell. (Cigar-boxt-tree), Myrtaceae as *Campomanesia xanthocarpa* (Mart.) O. Berg (guabiroba) and *Eugenia uniflora* L. (Brazil-cherry), besides the gymnosperm *Podocarpus lambertii* (Pinheiro-brabo) and the palm tree *Syagrus romanzoffiana* (Cham.) Glassman.

Other species occurring in association with *Araucaria angustifolia* are *Ocotea pulchella* (Nees & Mart.) Mez (Lauraceae), *Cinnamodendron dinisii* Schwanke (Canellaceae), *Moquiniastrum polymorphum* (Less.) G. Sancho (Asteraceae), *Matayba elaeagnoides* Radlk., *Cupania vernalis* Cambess. and *Allophylus edulis* (A. St.-Hil., Cambess. & A. Juss.) Radlk. (Sapindaceae), *Sloanea lasiocoma* K. Schum., (Elaeocarpaceae), *Luehea divaricata* Mart. (Malvaceae), *Handroanthus albus* (Cham.) Mattos (Bignoniaceae), *Myrsine coriacea* (Sw.) R. Br. and *Myrsine umbellata* Mart. (Primulaceae) and in secondary forests or riparian forests species such as *Ocotea puberula* (Rich.) Nees, in addition to *Mimosa scabrella* Benth. and other Fabaceae of the genus *Dalbergia, Inga, Lonchocarphus and Machaerium,* the Euphorbiaceae *Sapium glandulosum* (L.) Morong and *Gymnanthes klotzschiana* Müll. Arg., in addition to *Jacaranda puberula* Cham. and Asteraceae of the genus *Baccharis*, *Piptocarpha* and *Vernonanthura*, these taxa can vary widely due to the great successional state and due to the fragmentation that the forest is.

Between the nineteenth and twentieth centuries there was a great decrease in the areas of Mixed Ombrophilous Forest in Brazil, due to the logging of Parana-pine and associated species such as Brazilian-walnut and non-timber species such as the arborescent fern *Dicksonia sellowiana* Hook whose caudix was extensively used for the manufacture of pots for plants, with forested areas being gradually replaced by cities and mainly agricultural areas.

This period was marked by the expansion of the timber industry, in which the Mixed Ombrophilous Forest constituted the most significant area for this purpose in the Brazil's South. Between the 1950s and the 1960s, the main period of the timber cycle marked the sector development for more than a century of exploitation and led to the opening of areas for agriculture, livestock and, mainly, the timber industry in the mid of the twentieth century, with the so-called "Parana-pine cycle" in which the species wood represented 90% of all exports of the sector by the country.

The Ombrophilous Mixed Forest today is distributed in a mosaic of forest fragments, leaving only about 12% of its original area in Brazilian territory, with very sharp disjunction and areas most often smaller than 50 hectares. These remnants are frequently subject to negative effects of forest fragmentation, especially the edge effect, which is a more transforming and detrimental effect to these fragments, because with fragmentation there is a great correlation between the opening of areas, fragmentation and the edge effect, which only tends to increase.

In the state of Paraná, historically the site of greatest occupation of this forest, it is estimated that about 33% of the original area in which this physiognomy dominated remains. However, this volume unfolds into a mosaic rather small and disjoint fragments, with most of the vegetation occurring under the secondary initial or intermediate succession state.

Although this forest is highly endangered, forest restoration programs are scarce and lack scientific information on how the ecological processes occur in it, such as competition, seed facilitation and dispersal, and how to influence the natural growth of this forest community.

The recovery alternatives of these areas occur most often from reforestation projects, however, the forest has its own mechanisms of recovery and maintenance of its diversity, such as natural regeneration mechanisms, which include rainfall of seeds, soil seed bank and seedling bank.

The success of seed germination and its time in the soil depends on physiological factors (germination, dormancy and viability) and environmental factors (humidity, temperature, luminosity, presence of seed predators and pathogens) and the bank of these seeds have transient characteristics, being presented in two types: with seeds of short life, that do not present dormancy and germinate, generally, in the same year in which they were dispersed; or persistent, represented by dormant seeds that remain viable in the soil for more than one year.

Healthy seeds and seedlings are a prerequisite for any effort and success of a forest restoration activity. The importance of studies on longevity of forest species seeds, such as characteristics of seed dormancy and the remarkable diversity of patterns observed in relation to this factor. Therefore, the search for methods to perform pre-treatments to overcome seed dormancy without affecting the vigor of post-germinating seedlings is necessary, whereas studies of this approach relating to Mixed Ombrophilous Forest are scarce.

2. Overcoming Dormancy of Seeds

The dormancy is considered the seed inability to germinate during a certain period of time, under combinations of environmental conditions that would be favorable to germination as soon as the seeds overcome dormancy (BASKIN and BASKIN, 2004). Thus, in dormancy occurs the performance of genetically programmed mechanisms, resulting from the induction imposed by the combination of environmental conditions. The dormancy can be classified as:

- *Primary*: with systematic occurrence, does not depend on the region or year and presents variable intensity. It happens in the maturation period, being genetically programmed.
- *Secondary*: occurs sporadically, after maturity or dispersion, due to the environmental conditions. In this case, initially, the seeds were not dormant or had overcome dormancy.

The dormancy may be integumentary or exogenous and embryonic or endogenous, occurring independently of each other or simultaneously in the same seed. The dormancy can be physiological, integumentary, morphological, chemical, combined, in addition to slow or delayed germination.

- *Physiological dormancy*: it is caused by several factors inherent to physiological processes and by the absence of essences for germination, even the embryo being well developed. It can also be caused by some seed or fruit compound that block germination.

- *Physical dormancy*: it is the most common form of dormancy. It is related to the external parts of the fruit or seed, to the entrance of liquids or gases that promote germination. Resistance can be physical, when the integument is impermeable, or mechanical, when the fruit stiffness does not facilitate tegument rupture.
- *Morphological dormancy*: it occurs due to embryo immaturity, which is underdeveloped, thus requiring special conditions so that it can complete its development.
- *Chemical dormancy*: it is caused by inhibitory factors in the pericarp.
- *Combined dormancy*: it is caused by one or more factors that may be present blocking germination.
- *Slow or delayed germination*: although it is not considered a type of dormancy, the delay of 12 to 30 months to start the germination of a seed, besides irregularity in germination period, characterize as a specific type of dormancy.

Naturally, seeds overcome dormancy at the ideal moment by mechanisms of the seed itself or the environment, ending its cycle. As with dormant seeds that feed on animals and are later eliminated by feces are able to germinate. Another element is the cold, which in the winter period activates the production of substances that stimulate the embryo development. In addition to these elements, rain can also eliminate substances that may inhibit germination.

In Mixed Ombrophilous Forest, seeds of several forest species present dormancy (Table 1) and the majority of seeds are orthodox, having physical dormancy, zoocoric dispersion, and the majority dispersed by birds. In addition, most of them have low to moderate germination, probably because of seed type and dormancy. In this sense, recalcitrant seeds lose viability faster than orthodox ones, because of high humidity.

Table 1. List of seed characteristics of main forest species of the Mixed Ombrophilous Forest that present dormancy

Family/Species	Dispersion syndrome	Seed	Dormancy	Germination
Anacardiaceae				
Lithrea brasiliensis Marchand	Zoocoric	Orthodox	Physical	Low
Aquifoliaceae				
Ilex paraguariensis A.St.-Hil.	Zoocoric	Orthodox	Physical	Low
Araliaceae				
Schefflera morototoni (Aubl.) Maguire et al.	Zoocoric	Orthodox	Physical	Low
Araucariaceae				
Araucaria angustifolia (Bertol.) Kuntze	Zoocoric/Barocoric	Recalcitrant	Slow germination	High
Cannabaleaceae				
Trema micrantha (L.) Blume	Zoocoric	Orthodox	Physical	Low
Fabaceae				
Dalbergia brasiliensis Vogel	Anemocoric	Orthodox	Slow germination	Intermediary
Erythrina falcata Benth.	Autocoric	Orthodox	Physical	High
Mimosa scabrella Benth.	Autocoric	Orthodox	Physical	High
Ormosia arborea (Vell.) Harms	Autocoric/ Zoocoric	Orthodox	Physical	Intermediary
Senna multijuga (Rich.) H. S. Irwin & Barneby	Anemocoric	Orthodox	Physical	High
Lamiaceae				
Vitex megapotamica (Spreng.) Moldenke	Anemocoric	Orthodox	Physical	Low
Lauraceae				
Cryptocaria aschersoniana Mez	Zoocoric	Recalcitrant	Physical	Low
Ocotea odorifera (Vell.) Rohwer	Zoocoric	Recalcitrant	Physical	Intermediary

Family/Species	Dispersion syndrome	Seed	Dormancy	Germination
Ocotea porosa (Nees & Mart.) Barroso	Zoocoric	Recalcitrant	Physical	Intermediary
Ocotea puberula (Rich.) Nees	Zoocoric	Recalcitrant	Physical	Intermediary
Podocarpaceae				
Podocarpus lambertii Klotzsch ex Endl.	Zoocoric	-	Physical	Intermediary
Primulaceae				
Myrsine coriacea (Sw.) R. Br. ex Roem. & Schult.	Zoocoric	Orthodox	Physiological	Low
Proteaceae				
Roupala montana (Aubl.)	Anemocoric	Orthodox	Physical	High
Rosaceae				
Prunus myrtifolia (L.) Urb.	Zoocoric	Orthodox	Physiological	-
Styracaceae				
Styrax leprosus Hook. & Arn.	Zoocoric	Orthodox	Physical	High
Winteraceae				
Drimys brasiliensis Miers	Zoocoric	Orthodox	Physiological	Low

Table 2. Techniques used to overcome dormancy of forest species

Technique	Method
Chemical scarification	The seeds are immersed in substances that promote abrasion, aiming the integument corrosion. For this, sulfuric acid is used in a variable period according to the species, at a temperature of 19 °C to 25 °C. However, hydrogen peroxide and muriatic acid can also be used due to the toxicity of sulfuric acid.
Mechanical scarification	It is a technique used to overcome physical dormancy. The seed is subjected to a scraping or abrasion process so that the integumentary structure is abraded, making the structure permeable to the entrance of water and oxygen. In this procedure sandpaper is used, with wear intensity varying according to the species.
Immersion in running water	It is based on washing and removing chemicals that block germination. For this, the seeds are immersed in water for a period between 12 and 48 hours, with continuous flow.
Immersion in cold water	It is applied in species that do not present dormancy, however, the germination is slow, due to transport or to storage period and the seeds are dehydrated. In this technique, the seeds are immersed in water at room temperature for 24 hours.
Immersion in hot water or thermal shock	In this technique the seeds are immersed in hot water at recommended temperature and period for the species and then immersed in water at room temperature for a variable period.
Temperature switches	It is performed with seed placement to germinate in humid substrate with temperatures of 20 °C during the course of 8 hours and 30 °C or 35 °C, depending on the species, for 16 hours. This procedure can be carried out in a germination oven, under controlled temperature.
Cold Stratification	This method is used when the embryo is underdeveloped or immature. The seeds are submitted to low temperatures, between 5 °C and 10 °C, in order to favor the production of a growth hormone.
Overcoming combined dormancy	This technique is applied to seeds that present integumentary and embryonic dormancy. The seeds are submitted to two processes of dormancy overcoming, starting with the overcoming of integumentary dormancy.

Table 3. Techniques used to overcome the dormancy of seeds of the Mixed Ombrophilous Forest species

Species	Treatment	Reference
Araucaria angustifolia (Bertol.) Kuntze	Soak in water at room temperature for 24-48 hours.	Carvalho (2003)
Cryptocarya aschersoniana Mez	Mechanical scarification with the tegument seed cracking with pliers or hammer.	Fowler and Bianchetti (2000)
Dalbergia brasiliensis Vogel	Immersion in cold water for 48 hours.	Carvalho (1994)
Drimys brasiliensis Miers	Stratification in wet environment for 60 days.	Abreu (2005)
Erythrina falcata Benth.	Immersion in water at 80 °C, keeping the seeds in water until it reaches room temperature and for another 24 hours. Another alternative is the immersion of the seeds in water at room temperature for 48 hours.	David et al. (1995)
Ilex paraguariensis A. St-Hil.	Stratification in wet sand for 150 days.	Fowler and Bianchetti (2000)
Lithrea brasiliensis Marchand	Immersion in hot water at 80 °C, outside heating, for 3 minutes.	Medeiros and Abreu (2005)
Mimosa scabrella Benth.	Immersion of seeds in hot water at 80 °C, allowing them to cool and keeping them in the water at room temperature for 18 hours. Another option is to immerse in water at a temperature of 70 °C for 5 minutes.	Carvalho (1994); Floriano (2004)
Myrsine coriacea (Sw.) R. Br.	Immersion in water at alternating temperatures, 20 °C for 12 hours and 30 °C for 12 hours. It is also possible to proceed with stratification in wet sand for 30 days.	Carvalho (1994)
Ocotea odorifera (Vell.) Rohwer	Double dormancy: integumentary and embryo. Chemical scarification in sulfuric acid for 5 minutes, associated with stratification in wet sand for 60 days, with only one seed layer.	Carvalho (2003)

Table 3. (Continued)

Species	Treatment	Reference
Ocotea porosa (Nees & Mart.) Barroso	Mechanical scarification followed by stratification in sand or damp sawdust for 60 to 120 days. Another practical option is the use of solar scarification, placing the wet seeds in a place under direct sunlight. After drying, the tegument breaks easily and the seeds can be sown.	Carvalho (1994); Fowler and Bianchetti (2000)
Ocotea puberula (Rich.) Nees	Mechanical scarification followed by stratification in wet sand for 60 to 120 days.	Fossati (2007)
Ormosia arborea (Vell.) Harms	Mechanical scarification or immersion in water at a temperature of 80 °C to 90 °C for up to 48 hours.	Figliolia and Crestana (1995)
Podocarpus lambertii Klotzsch ex Endl.	Scarification with partial removal of the epimerus which involves the seed.	Carvalho (1994)
Prunus myrtifolia (L.) Urb.	Place the seeds to germinate in a shaded area (60% to 70% shade). The ideal germination temperature is between 20 °C and 25 °C constant. Another option is to pre-germinate in the dark at 35 °C.	Fossati (2007)
Roupala montana Aubl.	Immersion in water at room temperature for 24-48 hours.	Carvalho (1994)
Schefflera morototoni (Aubl.) Maguire et al.	Immersion in water at room temperature for 12 hours or immersion of the seeds in water at 65 °C, outside the heating, and then kept for 12 hours in the same water.	Carvalho (1994)
Senna multijuga (Rich.) H. S. Irwin & Barneby	Immersion of seeds in hot water at 100 °C (outside heating), leaving them soaked in water for 24 to 48 hours.	Carvalho (1994)
Styrax leprosus Hook. & Arn.	Mechanical scarification for two seconds using emery.	Fowler and Bianchetti (2000)
Syagrus romanzoffiana (Cham.) Glassman	Immersion in water at room temperature for 96 hours.	Fowler and Bianchetti (2000)
Trema micrantha (L.) Blume	Immersion in water at 50 °C, outside heating, for 5 minutes.	Floriano (2004)
Vitex megapotamica (Spreng.) Moldenke	It is not necessary. However, immersion in water at room temperature for 24 hours is recommended for imbibition.	Carvalho (2006)
Zanthoxylum rhoifolium Lam.	Scarification in concentrated sulfuric acid for 5 minutes.	Carvalho (2006)

Several techniques are used to overcome dormancy of forest seeds. Among these techniques, chemical and mechanical scarification are the most important methods applied in seeds of the Mixed Ombrophilous Forest species, due to high percentage of seeds with physical dormancy. In addition, some seeds require pre-germination treatment, even if they do not present dormancy. The procedure is performed to accelerate germination process. Table 2 shows the most important techniques for overcoming dormancy.

Actually, dormancy mechanism represents a beneficial mechanism, with the distribution of germination over time, increasing the species survival. Thus, the dispersion of seeds occurs in different stages of dormancy, causing the phenomenon known as heteromorphism or polymorphism, which directly influences the process of succession. The variations are characterized mainly by morphological characteristics of the seeds integument, generated by the environment; in addition to genetic factors.

Table 3 details the main techniques used to overcome seed dormancy of some species of the Mixed Ombrophilous Forest. It is verified that the majority of species that present dormancy belongs to the families Fabaceae and Lauraceae, mainly the species of the genus *Ocotea*, that are present in the forest upper stratum. In this sense, due to the majority of seeds presenting physical dormancy, it is verified that the most applied method is mechanical scarification, which consists in immersing the seeds in hot water or using acids.

Although some seeds of the Mixed Ombrophilous Forest species do not present dormancy, their viability period is quite reduced, because they are recalcitrant, that is, at the point of physiological maturity, there is a reduction of water content of the seeds. Most seeds of the species Myrtaceae and *Araucaria angustifolia* are classified as recalcitrant.

FINAL CONSIDERATIONS

In view of the great fragmentation and degradation of the Mixed Ombrophilous Forest, resulting in secondary forests, due to the total conversion in agricultural and urban areas, there is an urgency in conservation and restoration actions to avoid its depletion in the next centuries.

Forest recovery is most often based on techniques such as conducting natural regeneration by seed bank and induction of seed rain or through clearings and model modules that accelerate forest succession, as well as planting seedlings and enriching species by nucleation, which require basic knowledge on overcoming dormancy of the seeds and maximize germinative potential of these.

The main objective of this chapter was to present information regarding the overcoming of seeds dormancy of species from the Mixed Ombrophilous Forest, generating an important tool for better decision in management of forest recovery techniques.

It is also hoped that the current situation of the Mixed Ombrophilous Forest degradation in South of Brazil, which presents a rich biological diversity, will arouse interest of scientists, generating direct and indirect contributions for conservation and recovery of this very particular forest ecosystem.

REFERENCES

Abreu, D. C. A. Germination and morphological characterization of *Allophylus edulis* (St. Hil.) Radlk. and *Drimys brasiliensis* Miers [dissertation][internet]. Curitiba: Universidade Federal do Paraná; 2013. [cited 2018 June 02]. Available from: http://hdl.handle.net/1884/25948.

Accioly, P. *Mapping of remaining arboreal plants in the state of Paraná and elaboration of a geographic information system for the purpose of environmental analysis of the state* [thesis][internet]. Curitiba: Universidade Federal do Paraná; 2013. [cited 2018 May 30]. Available from: http:// www.floresta.ufpr.br/ defesas/ pdf_dr/ 2013/ t336_0376-D.pdf

Araucaria angustifolia in red List of the Brazilian flora version 2012. [internet]: *CNCFlora.* [cited 2018 May 30]. Available from: http://conifersociety.org/conifers/conifer/araucaria/angustifolia.

Araucaria angustifolia: Paraná Pine [internet]: Conifer Society. [cited 2018 May 30]. Available from: http://conifersociety.org/conifers/conifer/araucaria/angustifolia.

Auler, N. M. F., Reis, M. S. D., Guerra, M. P. and Nodari, R. O. The genetics and conservation of *Araucaria angustifolia*: Genetic structure and diversity of natural populations by means of non-adaptive variation in the state of Santa Catarina, Brazil. *Genetics and molecular biology*; 2002, 25(3): 329-338.

Backes, A. Climatic conditioning and geographic distribution of *Araucaria angustifolia* (Bertol.) Kuntze in Brazil - II. Researches. *Botânica*, 1999.; 49: 31-51.

Backes, A. Determination of the age and natural regeneration of a population of *Araucaria angustifolia* (Bertol.) Kuntze in a forest stand located in the city of Caxias do Sul, RS, Brazil. *Iheringia - Série Botânica*; 2001, 56(5): 115-130.

Backes, A. Current geographical distribution of the Araucaria Forest: climatic conditioning. In: Fonseca, C. R., Souza, A. F., Leal-Zanchet, A. M., Dutra, T. L., Backes, A., Ganade, G. *Forest with Araucaria. Ecology, conservation and sustainable development.*

Baskin, J. M. and Baskin, C. C. A classification system for seed dormancy. *Seed Science Research;* 2004, 14:1-16.

Behling, H. and Pillar, V. D. P. Late Quaternary vegetation, biodiversity and fire dynamics on the southern Brazilian highland and their implication for conservation and management of modern Araucaria forest and grassland ecosystems. *Philosophical Transactions of the Royal Society B: Biological Sciences*; 2007, 362(1478): 243-251.

Carvalho, P. E. R. *Brazilian forest species: silvicultural recommendations, potentialities and use of wood* Brasília-DF: EMBRAPA-CNPF/ EMBRAPASPI; 1994.

Carvalho, P. E. R. *Brazilian Tree Species* Volume 1. Brasília, DF: Embrapa Informação Tecnológica; Colombo: Embrapa Florestas; 2003.

Carvalho, P. E. R. *Brazilian Tree Species* Volume 2. Brasília, DF: Embrapa Informação Tecnológica; Colombo: Embrapa Florestas; 2006.

Churchill, S. P., Balslev, H., Forero, E. and Luteyn, J. L. *Biodiversity and conservation of neotropical montane forests*. New York: The New York Botanical Garden; 1995.

Davide, A. C., Faria, J. M. R. and Botelho, S. A. *Propagation of forest species*. Belo. Horizonte: CEMIG/UFLA/FAEPE; 1995.

Figliolia, M. B. and Crestana, C. S. M. Dormancy release in seeds of *Ormosia arborea* Arms. Revista do Instituto Florestal; 1995, 7(2): 259-65.

Floriano, E. P. *germination and dormancy of forest seeds* (*Germinação e dormência de sementes florestais*). Santa Rosa: ANORGS; 2004.

Fossati, L. C. *Ecophysiology of seed germination in populations of Ocotea puberula (Rich.) Ness, Prunnus sellowii Koehne and Piptocharpa angustifolia Dusén Ex Malme*. [thesis][internet]. Curitiba: Universidade Federal do Paraná; 2007. [cited 2018 May 28]. Available from: http://hdl.handle.net/1884/14584.

Fowler, J. A. P. and Bianchetti, A. *Dormancy in forest seeds*. Colombo: EMBRAPA-Florestas; 2000.

Garwood, N. C. Tropical soil seed banks: a review. In: Leck MA, Parker VT, Simpson RL. *Ecology of soil seed banks*. San Diego: Academic Press; 1989. P. 149-209.

Hueck, K. *The forests of South America: Ecology, composition and economic importance*. Brasília/São Paulo: Editora UnB/Polígono; 1972.

Hueck, K. *Die Wälder Südamerikas* [*The forests of South America*]. Stuttgart: Fischer; 1966.

Ibge – Instituto Brasileiro de Geografia e Estatística. *Technical Manual of the Brazilian vegetation*. 2.ed. Rio de Janeiro: 2012.

Ivanauskas, N. M. and Assis, M. C. Brazilian Forest Formations (Formações Florestais Brasileiras). In: Martins SVE. *Ecology of tropical forests in Brazil.* Viçosa: Editora UFV, 2009. P. 74-108.

Khurana, E. K. T. A. and Singh, J. S. Ecology of seed and seedling growth for conservation and restoration of tropical dry forest: a review. *Environmental Conservation*; 2001, 28(1): 39-52.

Laurance, W. F. and Vasconcelos, H. L. Ecological Consequences of Forest Fragmentation in the Amazon. *Oecologia Brasiliensis*; 2009, 13: 434-451.

Leite, P. F. and Klein, R. M. *Geography of Brazil*. vol. 2 Rio de Janeiro: IBGE. 1990.

Maack, R. *Physical geography of Paraná State*. Curitiba: Editora BADEP/ UFPR/IBPT; 1968.

Maack, R. *Physical geography of Paraná State*. Rio de Janeiro-RJ: José Olympio; 1981.

Medeiros, A. C. S. and De Abreu, D. C. A. *Instructions for testing the germination of native forest seeds of the Atlantic Forest*. Colombo: Embrapa Florestas; 2005.

Medeiros, J. D., Savi, M. and Brito, B. F. A. Selection of areas for the creation of conservation units in the Mixed Ombrophilous Forest. *Biotemas;* 2005, 18(2): 33-50.

Myers, N., Mittermeier, R. A., Mittermeier, C. G., Da Fonseca, G. A. B. and Kent, J. Biodiversity hotspots for conservation priorities. *Nature*. 2000; 403:853 858.

Nascimento, M. I., Poggiani, F., Durigan, G., Lemma, A. F. and Silva Filho, D. F. Effect of eucalyptus barrier on the containment of the border effect on subtropical forest fragment in the state of São Paulo, Brazil. *Scientia Forestalis*; 2010, 86(86): 191-203.

Oliveira-Filho, A. T., Budke, J. C., Jarenkow, J. A., Eisenlohr, P. V. and Neves, D. R. M. Delving into the variations in tree species composition and richness across South American subtropical Atlantic and Pampean forests. *Journal of Plant Ecology*; 2013. 8(3): 242-260.

Reis, A., Tres, D. R. and Scariot, E. C. Restoration in the Mixed Ombrophilous Forest through natural succession. *Pesquisa Florestal Brasileira*; 2007, 55:67-73.

Reitz, R., Klein, R. M. and Reis, A. *Santa Catarina State Wood Project*. Sellowia; 1978,28:1-320. Ribeirão Preto: Editora Holos; 2009. P. 39-44.

Ribeiro, M. C., Metzger, J. P., Martensen, A. C., Ponzoni, F. J. and Hirota M. M. The Brazilian Atlantic Forest: How much is left, and how is the remaining forest distributed? Implications for conservation. *Biological conservation;* 2009, 142(6): 1141-1153.

Ribeiro, M. C.,Martnsen, A. C., Metzger, J. P., Tabarelli, M., Scarano, F. and Fortin, M. J. The Brazilian Atlantic Forest: A Shrinking Biodiversity Hotspot. In: Zachos FE, Habel JC. *Biodiversity Hotspots: Distribution and Protection of Conservation Priority Areas.* Berlin: Springer-Verlag; 2011. P. 405-434.

Robinson, M., De Souza, J. G., Maezumi, S. Y, Cárdenas, M., Pessenda, L., Prufer, K. and Iriarte, J. Uncoupling human and climate drivers of late Holocene vegetation change in southern Brazil. *Scientific reports*; 2018, 8(7800): 1-10.

Roderjan, C., Curcio, G., Kuniyoshi, Y., Kirchner, F., Galvão, F., Kosera, C. and Accioly, A. *Diagnosis of the vegetation cover of the proposed area for the construction of the São Jeronimo SHP and the vegetative context of its surroundings*. Curitiba: FUPEF, 2003.

Souza, V. and Lorenzi, H. *Systematic botany: illustrated guide for the identification of families of native and exotic phanerogams in Brazil, based in APG III*. 3 ed. Nova Odessa: Plantarum; 2013.

Stefenon, V. M., Gailing, O. and Finkeldey, R. Phylogenetic relationship within genus *Araucaria* (Araucariaceae) assessed by means of AFLP fingerprints. *Silvae Genetica*; 2006, 55, 45–52.

Sccoti, M. S. V., Araujo, M. M., Wendler, C. F., Longhi, S. J. Mechanisms of natural regeneration in remnants of Deciduous Seasonal Forest. *Ciência Florestal*; 2011, 21(3): 459-472.

Zanini, L. and Ganade, G. Restoration of Araucaria forest: the role of perches, pioneer vegetation, and soil fertility. *Restoration Ecology*; 2005, 13(3): 507-514.

In: Plant Dormancy
Editor: Renato V. Botelho
ISBN: 978-1-53615-380-4

Chapter 9

Seed Physical Dormancy of Two Cultivars of Alfalfa (*Medicago sativa*) and White Clover (*Trifolium repens*): Similarities and Differences

Alberto A. Galussi*, María E. Moya, Felicia D. Zuriaga and Liliana R. Zimmermann
Faculty of Agricultural Sciences,
Entre Ríos State University, Entre Ríos, Argentina

Abstract

In forage legumes, seeds with physical dormancy (hard seeds) are not exceptional and during the germination test their embryos do not imbibe or germinate due to a physical impediment of their seed coats. This paper presents aspects related to physical dormancy studied in alfalfa (*Medicago sativa* L.) and white clover (*Trifolium repens* L.), and the results are

* Corresponding Author's E-mail: agalussi@arnet.com.ar; albegalu@gmail.com.

compared, showing similarities and differences between the samples of both species. The studies carried out in alfalfa seeds of cv. Baralfa 85 and white clover seeds of cv. NK Churrinche (2004 and 2006 harvests) analyzed the following aspects: weight and moisture content, passage of water to the embryo, viability, anatomical characteristics, determinations of structural substances, polyphenols, tannins, cutin and ions present in the testa of seeds with and without physical dormancy. The comparative analysis indicates that both species have physical dormancy with different levels of depth, which is overcome over time. The amount of seeds that break dormancy during the period when they are immersed in water is different between species. Physical dormancy is related to seed moisture content and modifications that occur in the seed coat, which differ between seeds without dormancy and hard seeds. The different degrees of depth of physical dormancy and the differences between species can be explained in terms of the characteristics of the testa: thicker cuticle and longer macrosclereids, thicker cell wall, and more conspicuous osteosclereids and in higher numbers, as well as accumulation of lignins, tannins, polyphenols and pectic substances, and the amount of cellulose and hemicellulose. The ions found can be related to the components of the cell wall in which they are found and the cementing properties of these substances. Physical dormancy is a mechanism that protects the embryo and, together with the low moisture content found in hard seeds, ensures longevity, prevents deterioration and predisposes most of the embryos to develop into normal seedlings. These characteristics of the seed coat prevent dehydration of the embryo and slow down exchanges with relative humidity of the environment, favoring good seed storage. A fraction of white clover seeds presented deeper dormancy than in alfalfa.

Keywords: forage legumes, physical dormancy, lignins, cell wall.

1. Introduction

1.1. Quality Parameters of Hard Seeds

A study conducted in 50 alfalfa fields showed that 90% of hard seeds were viable. Viability values ranged between 74% and 100%. In white clover, of 50 samples evaluated, the hypothesis that 100% of hard seeds are viable was accepted, with viability ranging between 84% and 100%. A hard seed, either with or without a viable embryo, does not germinate; it needs to overcome hardseededness and become permeable to be able to germinate and produce a normal seedling (Zimmermann et al. 1998). In a study that included 10 alfalfa samples and 10 white clover samples stored in temperate (20°C - 60% RH) and cold (4°C - 80% RH) environments, they exhibited different behavior in terms of longevity of hard seeds. On average, a 6% decrease in viability in hard seeds was reported. Although the storage capacity of the seed is improved in cold conditions, in all cases the hard seeds of these species maintained a high percentage of viability (Zimmermann et al. 2003).

1.2. Mathematical Expression of the Degree of Depth of Physical Dormancy

The initial step in seed germination is water uptake. This process also determines uniformity and speed of germination and establishment. In general, a small amount of water is absorbed during imbibition, likely less than two or three times the seed dry weight. In imbibed alfalfa seeds, the amount of water recorded was 93% of its dry weight (0.0028 g of water) (Galussi 2012). Water uptake in the seed usually occurs in three phases: an initial phase of rapid uptake, an intermediate phase in which water content remains almost constant, and a final phase of intense absorption related to cell elongation and radicle emergence (Bewley and Black 1986). Imbibition uniformity and speed depend on the morphology and permeability of the testas, the degree of contact of the seed with the humid substrate, and the differences in water potential (Besnier Romero 1989). In the system [water + testa], water in liquid state, atoms and molecules possess energy of motion, which is perceived as temperature; they are in constant motion and collide among them, but remain relatively close (*Kinetic molecular theory of matter*). The diameter of a water molecule is 3 A° and in liquid water, each water molecule forms hydrogen bonds with 3.4 other water molecules. In this system (water-seeds] the flow of water might be regarded as a passive process, meaning that water moves as a response to physical forces towards regions of low water potential or low free energy. When seeds are immersed in water, there is 100% of water, and although seeds are hygroscopic, the characteristics of the surface of permeable or impermeable testas (thin or thick waxy deposit, along with other cell components related to hydrophobia), respond to a range of behaviors, from absorption to rejection (unwettable hydrophobic surface).

Hydrophobia or wettability of a surface can be quantified by measuring the contact angles between droplets and the surface of interest. In surfaces of plant parts, this angle depends on the morphology of the epicuticular wax deposits (Ferrer Amorós 1997). Based on observations of the seed surface (Galussi et al. 2015; Galussi and Moya 2017), we hypothesize that in a harder (or more impermeable) seed, the contact angle increases, rendering the coat less wettable than a more permeable coat.

For the start of hydration, it might be necessary that waxes become wet or that pores or fissures be present in the cell layer (macrosclereids) of the testa so that water molecules gradually penetrate the coat, or that any enzyme activity starts so that the degradation process is triggered. In both species water entry occurs through the lens. The system seems static, but internally the entire system is moving (vibrational movement). Once water is in contact with the embryo cells, osmosis, i.e., water movement inside a plant cell, occurs, involving a combination of diffusion through the membrane bilayer and the flow of masses through the membrane pores. These pores are composed of aquaporins, integral proteins that form water selective channels through the membrane. Meyer et al. (2007) demonstrated that isolated membranes of both seed types (permeable and hard seeds) in soybean absorbed water at the same speed.

Using an empirical approach, seed analysts observe that not all seeds that are put to germinate uptake water and become swollen during the same period. Seeds that take longer time to become swollen due to water entry are supposed to be less permeable to water. Meyer et al. (2007) observed that when coats of hard soybean seeds were hydrated, they were permeable to water but their hydraulic conductivity was smaller than that of coats from permeable seeds by a factor of 5. Coats of both types of seeds showed weak osmometer properties, with low selectivity. Although that study confirmed that the cover is impermeable, it did not explain what factors confer impermeability to hard seeds.

If a group of alfalfa or white clover seeds belonging to the same lot—or the hard seed fraction of the sample—is under the same imbibition conditions, but not all seeds hydrate at the same time; how can this situation be expressed.

For that purpose, we modeled testa permeability to water to represent mathematically the time that seeds of a lot take to be completely hydrated. This experiment was performed using whole samples (data not shown) and hard seeds of alfalfa cv. Baralfa 85 and white clover cv. NK Churriche (Galussi et al. 2013[1], 2013[2]). A mathematical model is the accurate (mathematical) description of a real situation, such as physical, chemical or biological phenomena. It is a scheme, an equation, a diagram or a theory that represents mathematically a real situation. To construct the model, some assumptions are made and some simplifications of reality are considered. A model can be represented using relationships and/or functions. There are linear, quadratic, exponential models, among others. Model selection will depend on which model will best represent the real situation (Chapra and Canale 2004).

In our assay, after being immersed in water (3 repetitions of 200 hard seeds), seeds became hydrated in different proportions and periods; thus, different models were obtained for each assayed sample, either whole samples (hard and non-hard seeds) or only hard seed samples. This behavior would reflect the variability to water permeability between samples.

1.3. Causes of Differences in Degree of Depth of Physical Dormancy

Seed testa in legumes generally exhibits a common organization pattern. The outermost component is the single-layered epidermis composed of palisade macrosclereids, which are very tightly arranged and impregnated with chemical substances (quinones); the hilar region consists of two palisade layers. Below the epidermis is the hypodermis or columnar cells; it is composed of a single cell layer of sclerenchyma cells of very thick walls (osteosclereids). Then there is a deeper tissue, the lacunose parenchyma, composed of several flattened cell layers. Below this seed coat (testa) is the aleurone layer and the endosperm, then the embryo (Bewly and Black 1986; Esau 1987; Besnier Romero 1989). The presence of tegmen has been mentioned in some seeds (Marcos Filho 2005).

In species with impermeable testa, such as *Phaseolus lunatus*, the micropyle is blocked, the fissure of the hilum is closed (acting as a hygroscopic valve) and the strophiole remains intact. In non-hard legume seeds, water enters mainly through the strophiole (lens), a weak point in the palisade cell layers where fissures that allow water penetration develop (Simmons and Riedl 1968; Besnier Romero 1989). Hard and non-hard seeds do not vary in the anatomy of the testa (Nikolaeva 1980; Martens et al. 1995); however, Harris (1987) detected variations in the content of cutin (in the hilar region) and tannin, differences in porosity, and more pronounced light line in a soybean variety that contained hard seeds. However, its light line and the cuticle would not play any role in testa impermeability (Besnier Romero 1989; Boesewinkel and Bouman 1982).

Perissé and Planchuelo (2004) studied the testa and its relationship with water entry in lupine seeds; they found that the lens and pores found in this zone were the main path for water entry during imbibition. In hard seeds of annual legumes of the genera *Medicago* and *Trifolium*, Zeng et al. (2005) found differences in seed coat thickness (cuticle and outer layer) and profile (size and more or less compact cell arrangement).

In *Atriplex cordobensis*, Argüello et al. (1999) found epidermal and subepidermal cell layers with lignin accumulation in bracts, which the authors associated with water impermeability in seeds of that species, as well as the presence of Na, Cl and saponins, which partially inhibit germination. Furthermore, Alvarez et al. (1997) and Capeleti et al. (2005) related lignin content to resistance to mechanical damage in soybean cultivars. In some species of *Fabaceae* with physical dormancy, it has been mentioned that the lens is absent (Gunn 1984, 1991).

Which layer and which components are related to water impermeability are controversial issues and seem to vary with the study species. Noticeably, reports on seed of extreme longevity refer mainly to taxa with macrosclereid or osteosclereid cells in *Fabaceae* and *Malvaceae* (Mc Kee et al, 1977; Boesewinkel and Bouman 1982; Bewly and Black 1986; Martens et al. 1995).

The plant cell wall is composed of middle lamella, primary wall and in some cases, secondary wall. The cell wall is composed of the protoplast and is involved in nutrition, digestion, osmoregulation, growth, reproduction, intercellular communication and defense mechanisms. A typical primary cell (1-3 μm) can contain between 9-25% of cellulose, 25-50% of hemicellulose, and 10-35% of peptic substances, and about 10% of proteins. The secondary wall (approximately 4 μm thick) is composed of between 41-45% of cellulose, 30% of hemicellulose and in some cases 22-28% of lignin. When a cell that forms a secondary wall stops elongating, the lignin is deposited first in the middle lamella (composed of pectic substances), then in the existing primary wall and, finally, in the developing secondary wall (Salisbury and Ross 1992, Taiz and Zeiger 1998).

At maturity, in the seed testa of alfalfa and white clover, the outermost epidermal cell layer is the sclerenchyma tissue (macrosclereids) and, as the second cell layer (osteosclereids), the secondary wall is lignified. Lignification confers tissue with mechanical characteristics; indeed, the cell wall, depending on the distribution of cellulose microfibrils and matrix reinforcement of lignin, has a behavior similar to that of concrete, resistant to both tensions or rupture forces and compressions or crushing.

In alfalfa, Faner (2007) mentions the following values, according to the phenological stage: leaf and pre-flowering stage: lignin, 3.34% and 7%; hemicellulose, 4.07% and 8%; and cellulose 16.57% and 23%, respectively. In soybean seed coat, Alvarez et al. (1997) reported 6.2% lignin.

Testa hardness elicits several responses and their effects are diverse. In some legume species, cuticle wax present in the hard testa plays an important role in water impermeability. However, other studies suggested that the main barrier to water entry is posed by osteosclereids, since only when these are perforated, most of the seeds start to imbibe water (McKee et al. 1977; Bewly and Black 1986).

Peihua et al. (1992) studied structure and composition of chemical compounds of the seed coats of some species, such as *Ipomoea batatas* Lam., *Picea koraiensis* Nakai, *Dolichos dablab* L., *Sesbania cannabina* (Retz.) Pers., *Panaz quinquefolius* L., *Apium graveolens* L., *Nelummbo nucifera* Gaertn. and *Zea mays* L.. The authors found that K, Ca and Mg content is generally in the intermediate testa layer and that K and Ca content of the seed coat of different plants was positively correlated with the presence of hardness. Furthermore, Jose et al. (1986) found that Mg and other positive ions play a role in hardening of the seed coat. Some phenolic compounds found in hard testas would also play an important protecting role (Yassen et al. 1994).

In the plant cell, minerals are present as oxides and carbonates of Na, K, Ca, Mg, Al, Fe, Si, S, and P, mainly Ca and Mg as carbonates and silicates. They are embedded in several cell walls and in protoplasm or vacuoles of the cell in form of crystals (Sandoval et al. 2005). Calcium can act as a structural component of cell walls and membranes, and as cofactor of several enzymes. Calcium has been traditionally associated with the structure of the cell wall of which it is part as calcium pectate located in the middle lamella. Calcium has also been related to the cell membrane function (Garzón López 2003; Sandoval et al. 2005). Zeng et al. (2005) mentioned that pectins and hemicelluloses present in the walls become very hard at cell maturity in the presence of oxygen. As with calcium, magnesium can be found in plants as a structural element or as an enzyme cofactor. It is also a component of pectic substances contained in fruits and plays a role in resistance mechanisms of tissues. The main role of potassium is that of activating several enzymes.

The presence of aluminum might also trigger callose production. An increase in cytosolic calcium would be a prerequisite for the formation of callose (1,3, β glucan beta-glucan), whose production can be related to any type (physical, chemical, biochemical or biological) of stress affecting plants (Kauss, 1985). Several proteins containing Cu play a key role in processes such as lignification, photosynthesis, respiration and superoxide radical detoxification (Kirkby and Römheld 2007).

2. Quality Attributes of Hard Seeds. Comparative Analysis

Comparative studies of hard seeds of alfalfa and white clover were conducted using the results of analyses performed in samples from two harvests (2004 and 2006) from the same production fields in Argentina (Galussi et al. 2013a, 2013b, 2015; Galussi & Moya 2017). The hard seed fraction was separated by putting working seed samples in water during 22 hours. Those seeds that did not imbibe were considered hard seeds (i.e., seeds with physical dormancy) (ISTA 2003, 2007).

2.1. Weight of Seeds

Weight of 1000 hard seeds of alfalfa and white clover is slightly lower than weight of seeds of the whole samples (mostly composed of non-hard seeds and hard seeds, alfalfa contained: 16% and 3%, and white clover 11% and 7% of hard seeds, for 2004 and 2006 harvests, respectively). Weight of white clover seeds is lower than weight of alfalfa seeds, with the latter ranging between 2.49 and 2.21 g and the former, between 0.649 and 0.626 g, whereas weight of whole samples ranged between 2.59 and 2.49 g (alfalfa) and 0.657 and 0.671 g (white clover) for the 2004 and 2006 harvests, respectively. The fact that weight values of hard seeds are similar to those of the global samples would reveal that hard seeds reach physiological maturity and the maximum dry weight. This means, dormancy would not occur before that stage is reached, and therefore maximum dry weight similar to that of non-dormant seeds would be reached. The slight difference from the weight of the global seed sample might be due to the lower water content of hard seeds.

2.2. Moisture Content of Seeds

Moisture content in hard seeds of white clover was lower than that of alfalfa hard seeds (about 1% lower on average). In both species, moisture content of hard seeds was between 3.5 and 4% lower than that of whole samples (of the same lot), according to data from samples taken in two harvests. Moisture content values found in hard seeds were 6.1 and 7.2% in alfalfa and 5.9 and 6.4% in white clover for 2004 and 2006 harvests, respectively; these values correspond to moisture content of seeds that are in equilibrium with 30% to 45% of relative humidity (RH) in the environment (Harrington 1972; Aguirre and Peske 1988; Moravec, et al. 2008). These results suggest that, after physiological maturity, days of low RH would promote the formation of hard seeds, as mentioned by Harper (1991) for a green pea variety. Mean water content found in hard seeds (6-7%) would determine the presence of water type 1 in hard seeds. At this physiological level, metabolic activity is restricted, with no enzyme-mediated reactions taking place and with catabolic events being dominant (Vertucci and Leopold 1986; Parmmenter and Berjak 2000). Seeds of permeable testas would present not only type 1 water but also type 2 water, since its water content ranged between 8 and 9.5% in alfalfa and 9.2 and 9.9% in white clover (Galussi et al. 2013a, 2013b).

2.3. Water Entry into Seeds

The studies conducted to determine water entry into seeds with and without dormancy in alfalfa and white clover (Galussi et al. 2013a, 2013b) showed that in both non-dormant and hard seeds (seeds with physical dormancy) water enters through the lens. In those studies, water entry was clearly observed by staining, showing the sequential images of water entry over time. Other authors have also mentioned the lens as the entry site in leguminous seeds, but not specifically in alfalfa or white clover hard seeds (Besnier Romero 1989; Côme and Courbineau 1992; Perissé and Planchuelo 2004; Marcos Filho 2005). However, some alfalfa seeds have been found to have an additional entry site in the area of the coat above the hypocotyl-radicle axis, as indicated by an area of stained grooves, denoting water penetration. Anyway, in seeds with physical dormancy, in both species, hydration time is markedly different (much slower) from the one found in seeds with permeable coats; this difference shows, to some extent, the hydrophobic characteristic of hard seeds (Galussi et al. 2013a).

The results of the viability analysis of hard seeds that were compared for this work are those obtained from the germination assays (Galussi et al. 2013a, 2013b): in alfalfa of 16% of hard seeds (2004 harvest), 96% had viable embryos, and of 3% of hard seeds (2006 harvest), 100% had viable embryos. In white clover, of 11% of hard seeds (2004 harvest), 98% were viable, and of 7% of hard seeds (2006 harvest), 100% of embryos were viable. Non-viable embryos were attributed to damaged tissue of the radicle and the hypocotyl-radicle axis, as mentioned by Zimmermann et al. (1998, 2003). Hard seeds have high viability values, which may reach 100%, depending on the species, initial physiological quality, environment and storage period.

Dormant seeds of alfalfa and white clover seeds were found to have an appropriate moisture content, which provided them with good physiological quality. Experimental evidences confirm that in alfalfa and white clover seeds, dormancy is imposed by the water-impermeable seed testa. After a given period, seeds overcome dormancy, allowing water entry through the lens, with no differences in the process from non-dormant seeds.

Viability of dormant seeds is higher than that of non-dormant seeds. This phenomenon may be explained by the lower moisture content of the former seeds, which favors storage potential due to the decrease in peroxidative processes. Hard seeds of white clover had lower moisture content than alfalfa seeds, suggesting greater longevity of their embryos.

3. Mathematical Expression of the Degree of Depth of Physical Dormancy

The comparison of the results obtained in the published works (Galussi et al. 2013a, 2013b) is presented in Figure 1, showing the water permeability curves of the hard seeds of the tested samples through time. The samples of hard seeds of both species show different behavior in water permeability (different depth of physical dormancy).

In alfalfa, the saturation model assayed with hard seeds (2006 harvest) separated after the first 22 hours of immersion, and then put in water for 387 days (9,279 hours), is expressed mathematically by the minimum squares method

$$y = \frac{1392.9833775\, x^{0.4}}{584.30819 + x^{0.4}} \quad (1)$$

and is illustrated in Figure 1, with a Coefficient of determination $R^2 = 0.9956766$; where **y**: percentage of permeability, **x**: hours of immersion in water.

The permeability assay performed with hard seeds of white clover (2006 harvest) ended after 542 days (13,004 hours) of immersion in water. White clover seeds had a different behavior from that of alfalfa seeds, and the mathematical function was different. The permeability model of the hard fraction of this sample is expressed with the function

$$y = 9.96945 + 0.00628644\, x^{0.9} \quad (2)$$

The curve for alfalfa refers to a model fit similar to a homographic function, with a pronounced initial growth, and tending to become asymptotic over time for 100% of permeability. The white clover curve presents an almost linear model fit, with a constant but very slow growth over time, becoming stable at the end of the experiment, i.e., about 45% of permeability. Interpretation of both curves shows a notable difference in the degree of depth of physical dormancy between seeds (water permeability evolution).

After 8 days (192 hours), the proportion of seeds that hydrated and germinated to develop into normal seedlings was 11% in white clover and 20% in alfalfa. After 56 days (1400h) of immersion in water, the proportion of hard seeds that became permeable was 14% for white clover and 42% for alfalfa.

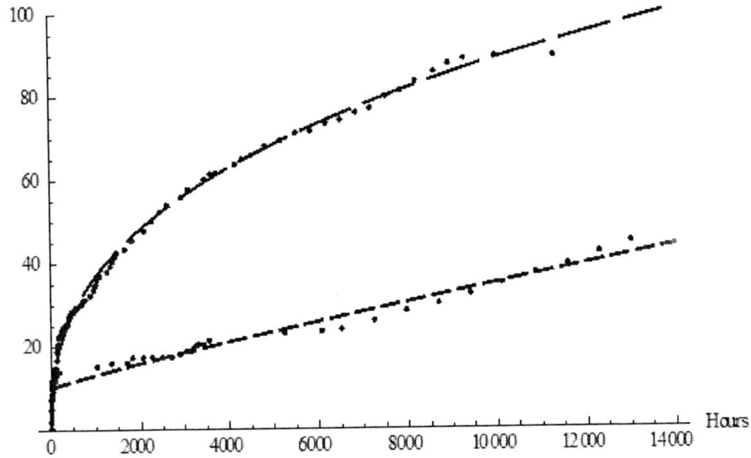

(a)

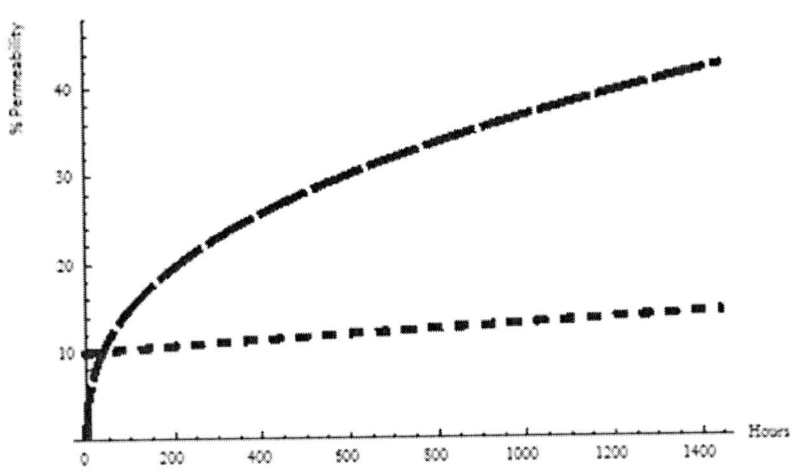

(b)

Figure 1. Evolution of water permeability of hard seeds of *M. sativa* cv. Baralfa 85 and *T. repens* cv. NK Churrinche ( ) 2006 harvest. (a) 0 – 14,000 hours period of water immersion (b) Same model showing the first 1,400 hours of immersion in water.

Seeds of alfalfa and white clover differ in their permeability, which may be explained with mathematical models. However, some general aspects can be stated. In hard alfalfa seeds, the curve represented by the mathematical model is asymptotic, indicating that finally all seeds become hydrated over a given period that can be estimated. In white clover, the curve is not asymptotic, suggesting that a given fraction of hard seeds (about 50%) had very deep dormancy, and with some differences among seeds, would finally become hydrated, but very gradually and slowly. A general and very important aspect for hard seeds that become hydrated is that in both species and in the assayed samples, after breaking dormancy these seeds develop into normal seedlings. This fact shows that dormancy mechanism is crucial for embryo protection.

If similar assays were performed with other samples, it is likely that each sample would have a different saturation permeability model; however, the degree of depth of physical dormancy was confirmed in these hard seed samples.

4. Weight of Metal Ions in the Testa (Alfalfa and White Clover)

The determination of ions (by weight) in the teguments of alfalfa cv. Baralfa 85 and white clover cv. NK Churrinche (2004 and 2006 harvests) were carried out in CERIDE, using the technique of Atomic Absorption Spectrometry (Analytical Methods-Perkin Elmer 1996).

The ions that were quantified were selected according to their participation in impermeability. Determinations of aluminum (Al), copper (Cu) and manganese (Mn) were made by Electrothermal Atomic absorption spectrometry under STPF conditions and calcium (Ca), magnesium (Mg), potassium (K), silica (Si) and sodium (Na) determinations were made by flame atomization. Samples (1-2 g each) of seed testas (alfalfa and white clover of the studied cultivars) of different permeability degrees were treated with microwave digestion in a Milestone START D digester. The method of calibration curve with certified aqueous standards was used. A Perkin Elmer Analyst 800 spectrometer was employed. Total phosphorus was determined using the 4500 – P C Vanadomolybdophosphoric colorimetric Acid method (1998). Samples were treated by microwave digestion in a Milestones ETHOS 900 digester. An Ultraviolet – Visible Perkin Elmer Lambda 40 Spectrometer was used. This technique quantifies ions in mg or µg, but not qualitatively as the analysis conducted by X Ray Fluorescence Spectrometry. Quantitative data of ions were analyzed by means of Principal Component Analysis (Di Rienzo et al. 2010).

Table 4.1. Quantity (average) of ions detected in testas of seeds of *Medicago sativa* cv. Baralfa 85 and *Trifolium repens* cv. NK Churrinche with or without dormancy

Ion determination (average)	Seed testas							
	M. sativa				*T. repens*			
	WT/D		W/D		WT/D		W/D	
	A	B	A	B	A	B	A	B
Aluminum (µg/g)	28-5	32	25	50	157	59	160	71
Calcium (mg/g)	12	12	12	14	13	10	14	11
Copper(µg/g)	12-5	13	16	20	12	15	16	15
Magnesium (mg/g)	1-4	1-1	1-3	1-5	0-98	1-6	0-99	0-9
Manganese (µg/g)	17	12	23	21	19	16	20	13
Potasium (mg/g)	0-30	0-23	0-45	0-40	0-31	0-39	0-25	0-27
Sodium (mg/g)	0-5	0-38	0-469	0-59	0-83	0-25	0-81	0-33
Total Phosphorus (mg/g)	0-65	0-7	0-7	0-5	1-0	1-6	0-8	1-2

WT/D: testas of seeds without physical dormancy; W/D: testas of seeds with physical dormancy. (A) 2004 harvest; (B) 2006 harvest.

The determined amounts of ions present varied between the years of harvest of the teguments (Table 4.1.). The evaluation by the Principal Component Analysis (data not shown), per year and together, shows, for some ions, a tendency of association according to the condition of the seed without or with dormancy.

Table 4.2. Characteristics of seeds and testas of *Medicago sativa* cv. Baralfa 85 and *Trifolium repens* cv. NK Churrinche, from two harvests (2004 and 2006).

Characters*	Dormant sedes		Non-dormant sedes	
Testa	Alfalfa	White Clover	Alfalfa	White clover
Thickness (μm)**	60	45-50	30	30 – 40
Cuticle (μm)**	5	3 – 5	1	1
Macrosclereids** (length) (μm)	40 – 50	30- 40	20 – 30	25 - 30
Macrosclereids** (width) (μm)	12.3 – 5.50	9.4x 3.4	9.17 – 5.50	2.7 x 1.8
Cell wall** (thickness in macrosclereids) (μm)	2.3	2.2	1.2	0.54
Osteosclereids** (length) (μm)	8 – 10	(5-8) x (15-18)	0 – 3	No halladas
Total polyphenols (%)	1.41	-	0.8	0.47
Condensed tannins (%)	0.63	0.29	0.48	0.16
Cell content	25.3	22.9	13	67.2
Cell Wall	74.7	77.1	87	32.1
Lignins (%)	7.7	5.5	5.2	1.3
Cellulose (%)	39.9	48.6	45.8	21.0
Hemicellulose (%)	20.7	16.1	60.4	6.9
Cutin	4.7	-	8.2	0.5
Ions***	> Mg, Mn, K, Na, Cu, Al	> Mg, Mn, Cu	< Mg, Mn, K, Na, Cu, Al	> Na, P, K

* Mean values of approximately 3g (3000) of testas of seeds with and without physical dormancy. ** Mean values of five seed testas with or without physical dormancy (after being immersed in water for 3000 hours). *** Number of ions determined in a sample of 1-2g of testa of seeds with or without physical dormancy shown in Table 4.1. (-) Not detected.

Data taken from Galussi et al. (2015); Galussi and Moya (2017)

In testas of alfalfa of 2004 harvest, dormancy was strongly associated with Mn, K and Cu, whereas permeability was associated with Na and Mg. Aluminum and Ca did not show a strong trend towards an association with any seed condition. In testas of 2006 harvest, dormancy condition was strongly associated with K, Mn, Mg, Na, Cu and Al, whereas permeability was associated with P. Ca and Na did not show a strong trend towards an association win any condition. The results differed between years.

In white clover, data obtained using this technique (Table 4.1.) show slight differences in quantities between seeds with and without dormancy; however, in testas of dormant seeds, there was a higher quantity of aluminum, calcium and copper; a lower quantity of potassium, and variable amounts of magnesium, manganese, sodium and phosphorus, depending on the harvest year. For the non-dormant condition, the associated elements were Na, P and K; Ca, Mg and Mn were present in testas of seeds with and without dormancy in similar proportions; therefore, there was no clear association with any condition.

Considering the possible association of the different aspects described above and physical dormancy, we postulated the hypothesis that the different degrees of physical dormancy in alfalfa and white clover seeds can be explained by anatomical, physiological, and chemical characteristics of the testas that determine water permeability (Galussi et al. 2015; Galussi and Moya, 2017). The results (Table 4.2.) provide evidences to accept the hypothesis in the two cultivars of both species. The comparative studies involved testas of non-dormant seeds (seeds that imbibed within 4 hours of immersion in water) and dormant seeds (seeds that did not imbibe after being immersed in water for 125 days -3000 hours). The values found in both species were not similar; however, in testas of dormant seeds of both species, cell layers were thicker and the proportion of cementing and water repelling substances was higher than in non-dormant seeds. Anatomical characteristics were determined in five seeds with different degrees of dormancy, whereas chemical components were measured in a sample (3 g) of hard seed testas (approximately 3000 testas) of each cultivar. These data represent specific anatomical values and averages of the chemical components, with possible variability among testas, which confers different levels of dormancy depth, as observed in the permeability curve obtained.

5. Anatomical, Physiological and Chemical Characteristics of the Testa Relative to Water Permeability

Dormant seeds of alfalfa and white clover have a thick cuticle all over the testa surface; macrosclereids are compact, longer and wider, with thicker cell walls in alfalfa than in white clover.

In testa of dormant seeds, there is a clear arrangement of abundant osteosclereids, which confers the testa with mechanical resistance. They are greater and more conspicuous in hard testas of white clover than in those of alfalfa.

Testas of hard seeds in both species had a higher amount of defense and hydrophobic components, such as polyphenols and tannins, lignins and pectates. White clover exhibited a thicker cell wall, which may explain the deeper physical dormancy than that of alfalfa.

The amounts of ions found in testas of dormant and non-dormant seeds did not show marked differences; however, ions are involved in cell physiological processes and would be related to the greater amount of cell wall components present and to the cementing and hydrophobic properties of these substances.

The seeds of the lot had different cellular arrangement and amounts of chemical compounds associated with water impermeability; therefore, they reflect the variation in the degree of permeability through mathematical models.

The effect of the treatment of hard seeds of alfalfa under humidity and low temperature conditions was observed in an experiment conducted by Galussi (2011). The seeds (4 repetitions of 100 hard seeds) placed between moist paper and at 10°C showed water permeability in 30, 50 and 77% at 8, 33 and 218 days, respectively. When immersed in water at a temperate temperature (20 ± 1°C), 30, 50 and 69% became permeable to water at 24, 95 and 218 days, respectively (Figure 2). In alfalfa seeds with physical dormancy placed on moist substrate at low temperatures, the dormancy period is shortened, since these conditions favor the elimination of the cuticle, as observed in the microphotographs taken with SEM (Figure 3) and the "softening" (possibly by enzymatic action) of the testa, allowing the entry of water in less time.

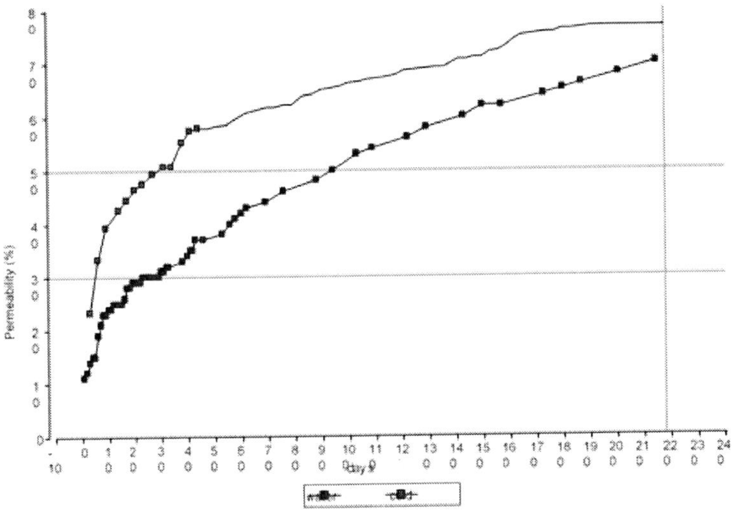

Figure 2. Seeds of *Medicago sativa* (alfalfa) cv. Baralta 85 (2006 harvest) with physical dormancy that became permeable after moist chilling or water immersion treatments. (Permeability %): percentage of permeable seeds; (days) days that seeds were between moist paper layers and in the dark at 10°C or immersed in water at room temperature and dark conditions.

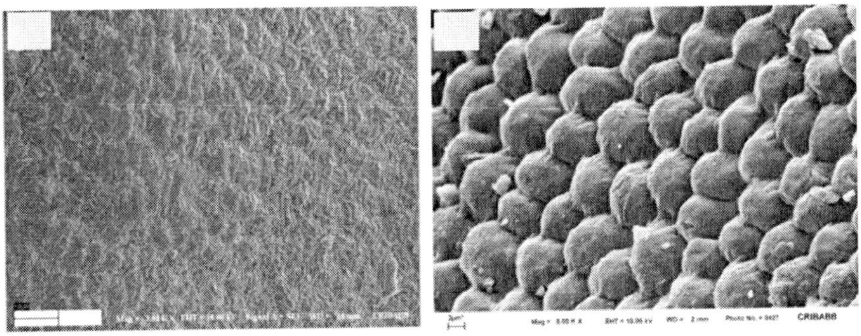

(a) (b)

Figure 3. Testa of seeds of *Medicago sativa* (alfalfa) cv. Baralfa 85, with physical dormancy, treated to induce dormancy breakage. (A) untreated seeds (scale bar: 10μm), (B) testa after moisture and cold treatment (scale bar: 2 μm).

Conclusion

The different behavior of alfalfa (*M. sativa*) and white clover (*T. repens*) seeds with physical dormancy is explained in terms of the different levels of dormancy depth, which are given by the anatomical, physiological and chemical characteristics of the seed testa. This hypothesis is confirmed based on the found evidences (Table 4.2.), which were demonstrated in the previous sections and from which the following conclusions are drawn.

- In both species only dormancy was evidenced, since the embryo starts germination once water entry and hydration occurs.
- In hard seeds without apparent physical damage (intact seeds), water enters through the lens area and extends to the proximal end of the lens (opposite to the radicular lobe), and then around the hilar rim, meaning that the lens is the natural site of water entry to the seed in both species.
- The different degrees of dormancy found in the seed population conditions water entry and may cause staggered germination over a long period, which would be longer in white clover than in alfalfa.
- The quantity of seeds with physical dormancy (hard seeds) can vary between samples; hard seeds show better physiological quality than non-dormant seeds.
- The low water content found in dormant seeds corresponds to the moisture content of seeds in equilibrium with low relative humidity (30% to 45%) in an environment.
- It can be assumed that, towards physiological maturity, when the embryo has high water content, a dehydration process starts until equilibrium with RH is reached; thus, the start of the dormancy mechanism avoids greater loss of water.
- In hard seeds, duration of phase 1 of germination was variable; the variability of hydraulic conductivity and dormancy depth can be expressed in mathematical models of water saturation. Seeds

considered dormant gradually became permeable as a function of time and depth of dormancy, showing that dormancy has a limited duration, which was longer in white clover.

Surface, thickness and structure of testas exhibited notable differences between dormant and non-dormant seeds. Testas of dormant seeds were thicker (more than two-fold), due to cuticle thickness and length of macrosclereids and osteosclereids. In alfalfa, testa is thicker than in white clover. Macrosclereids of testas differed between dormant and non-dormant seeds in cell size (up to more than two-fold in length) and cell wall thickness (about 2 μm thicker in seeds with physical dormancy). Abundant and conspicuous osteoscleroids were observed in hard seeds, mainly in white clover. This cellular arrangement in the testa of dormant seeds, along with structural components, provides the testa with more mechanical resistance.

Differences in the quantities of chemical components of the testa of dormant seeds versus the testa of non-dormant seeds were detected in both species.

Such differences are associated with impermeability/permeability to water. The Principal Components Analysis (Galussi et al. 2015; Galussi and Moya 2017) shows an association between impermeable testas (dormant seeds) and components present in greater proportion and that have hydrophobic and cementing properties, such as polyphenols, lignins, condensed tannins, pectic substances and low amount of the hydrophilous substances, cellulose and hemicellulose.

The great proportion of lignin found in dormant seeds and its roles, i.e., conferring more rigidity and resistance, regulating hydration of cellulose to favor wall elasticity, establishing an impermeable barrier to the cell, and forming a protection cover of the wall against the attack of physical, chemical and biological agents, show that lignin is the most important component of macrosclereids and osteosclereids associated with dormancy in alfalfa and white clover seeds.

The association with impermeability of ions that were found in great amount should be related to physiological processes in which each ion is involved, mainly in lignification, and to phenols and pectates.

The different cell arrangement and amounts of chemical components associated with water impermeability and hardness that are present in seeds of a lot cause different degrees of dormancy that can be represented through mathematical models of permeability.

Subjecting hard seeds to moist chilling reduces the cuticle layer, causing permeability of the testa over time. The continuous movement of the water molecules causes penetration, with the consequent moistening. Moreover, activity of enzymes (pectinase and hemicellulase) that might degrade the middle lamella, favoring water entry, would be triggered.

Physical dormancy of seeds is determined by several factors, and is summarized in the function: *Depth of Physical Dormancy* (3):

$$\boldsymbol{DPD = f}\ [\mathbf{Ef, SWC, St}] \tag{3}$$

$\mathbf{Ef} = f_1$ [RH, T°] $\mathbf{SWC} = f_2$ [Ef, G]
$\mathbf{St} = f_3$ [Cu, D, Cl]; $\mathbf{Cl} = f_{31}$ [M, O]; $\mathbf{M} = f_{311}$ [l, cw, cwt];
$\mathbf{O} = f_{312}$ [Si, Nu];
$\mathbf{CwM} = f_{3111}$ [Li, Tp, Ct, Ps, Ce, Hc]
SPD → MVE

Depth of Physical Dormancy (**DPD**) in alfalfa seeds is a function (***f***) of the Environmental Factors (**Ef**), seed water content (**SWC**) and the characteristics of the seed testa (**St**). The Environmental factor varies (f_1) with Relative Humidity (RH) and Temperature (T°). **SWC** depends on environmental factors, relative humidity and temperature, and the genotype (G). Modifications of the seed testa depend on the Cuticle (Cu), Depressions (D) and cell layers of the testa (Cl). Cell layers of the testa (Cl), the only integument found in alfalfa and white clover, is a function of Macrosclereids (M) and Osteosclereids (O). In turn, Macrosclereids vary as a function of length (l), cell width (w) and thickness of the cell wall (cwt). Osteosclereids in testas of dormant seeds are a function of size (Si) and Number (Nu) present. The hydrophobic characteristic of the Cell Wall of Macrosclereids (**CwM**) (the most important characteristic of dormant seeds) varies as a function of the proportion of its chemical components: higher contents of

Lignin (Li), Total polyphenols (Tp); Condensed tannins (Ct) and Pectic substances (Ps), and amount of hydrophilic substances -lignified Cellulose (Ce) and (Hc) lignified Hemicellulose. As a consequence of dormancy mechanism, Seeds with Physical Dormancy (SPD) contain Mostly Viable Embryos (→MVE).

Depth and limited duration of physical dormancy were observed in the analyzed seed lots. When does dormancy mechanism start? It has not been demonstrated, but it would end when moisture content found in hard seeds was reached. Seeds with deeper dormancy will take longer to become permeable than those with less deep dormancy. Most seeds with physical dormancy contain viable embryos and once permeability to water and other appropriate factors occur, normal seedlings usually develop.

The characteristics of the seed coat, the lower moisture content and viable embryos in hard seeds indicate that the mechanism of physical dormancy contributes to preserve the embryo during the time that the seeds are dormant.

REFERENCES

Aguirre R. and Peske, S. T. (1988). *Manual Para el Beneficio de Semillas*) [*Manual for the Benefit of Seeds*]. XII-15. Universidad del Estado de Mississippi. Centro Internacional de Agricultura Tropical (CIAT)., Cali, Colombia.

Alvarez, P. J. C., Kryzanowski, F. C., Mandarino, J. M. G., Franca Neto, J. B. (1997). Relationship between soybean seed coat lignin content and resistance to mechanical damage. *Seed Science and Technology*. 25 (2), 209-214.

Analytical Methods-Perkin Elmer. (1996). *Analytical Methods for Atomic Absorption Spectroscopy*. The Perkin-Elmer Corporation. United States of America. 300 pp.

Argüello, J. A.; Seisdedos, L.; Cid, M. V. and Aiazzi, M. T. (1999). Morphophisiological aspects of dormancy in the dispersal of *Atriplex cordobensis. ΦYTON International Journal of Experimental Botany*, 65: 35-41.

Baskin, J. M. and Baskin, C. C. (2004). A classification system for seed dormancy. *Seed Science Research*, 14, 1-6.

Besnier Romero, F. (1989). *Semillas, Biología y Tecnología* [*Seeds, Biology and Technology*]. Ed. Mundiprensa 637 pp.

Bewley, J. and Black, M. (1986). *Seeds. Physiology of Development and Germination*. Pleum Press. 367 pp.

Boesewinkel, F. D. and Bouman, F. (1982). The Seed: Structure (567-608). In: *Experimental Embryology of Vascular Plants*. Ed. Jori, B. M.

Capeleti, I., Ferrarese, M. L., Krzyzanowski and Ferrarese F. (2005). A new procedure for quantification of lignin in soybean (*Glycine max* (L.) Merrill) seed coat and their relationship with the resistence to mechanical damage. *Seed Science and Technology*. 33: 511-515.

Chapra, S. and R. Canale. (2004). *Métodos numéricos para ingenieros* [*Numerical Methods for Engineers*]. Ed. Mc. Graw Hill. Méjico, pp. 319–341.

Côme, D. and F. Courbineau. 1992. Environmental control of seed dormancy and germination. In: *Advances in the Science and Technology of Seeds*. Ed. Fu Jiarui and Khan, A. A. Science Press. Pp. 277-298.

Di Rienzo, J. A., Casanoves, F., Balzarini, M. G., Gonzalez, L., Tablada, M., Robledo, C. W. (2010). *InfoStat versión 2010.* Grupo InfoStat, FCA, Universidad Nacional de Córdoba, Argentina.

Esau, K. (1987). Anatomía de las Plantas con Semillas [Anatomy of Seed Plants]. pp. 423-430. *Hemisferio Sur*, Buenos Aires, Argentina.

Faner, C. (2007). La pastura de alfalfa como fuente de alimentación para cerdos en crecimiento y terminación [Alfalfa pasture as a source of food for growing and finishing pigs] *Boletín AACP* 01.10.07. In: web: www.produccion-animal.com.ar/ produccion_porcina/ 72-alfalfa.pdf (2009).

Ferrer Amorós, J. (1997). *Las células de los tejidos vegetales* [*The cells of plant tissues*]. Ed. Vedrá, S.L., 392 pp. Barcelona, España

Galussi, A. A. (2011) Estudios morfológicos y fisiológicos de la dormición de semillas de *Medicago sativa* L. (alfalfa) cv. Baralfa 85.

[Morphological and physiological studies of the seed dormancy of *Medicago sativa* L. (alfalfa) cv. Baralfa 85]. Tesis doctoral. Facultad de Ciencias Agropecuarias - Universidad Nacional de Córdoba, Argentina.

Galussi, A. A.; Argüello, J. A.; Moya, M. E.; Zuriaga, F. D.; Zimmermann, L. R. (2013a). Seed dormancy mechanism as a factor influencing seed physiological quality in alfalfa (*Medicago sativa* L.) cv. Baralfa 85. *Seed Science and Technology*, 41, 50-59.

Galussi, A. A.; Moya, M. E.; Zuriaga, F. D.; Zimmermann, L. R.; Basso R. (2013b). Effect of the dormancy mechanism in the quality of seeds of *Trifolium repens* L. (white clover) *Seed Technology*. Vol. 35 (2): 25-37.

Galussi A. A.; J. A. Argüello; M. M. Cerana; M. Maximino; M. E. Moya. (2015). Características anatómicas y químicas del tegumento seminal de *Medicago sativa* L. (alfalfa) cv. Baralfa 85 y su asociación con la dormición [Anatomical and chemical characteristics of the seed coat of *Medicago sativa* L. (alfalfa) cv. Baralfa 85 seeds and their association with seed dormancy]. *ΦYTON International Journal of Experimental Botany*, 84: 163-175.

Galussi A. A. and M. E. Moya. (2017). Anatomical and Chemical Insights into the White Clover (*Trifolium repens* L.) Seed Coat Associated to Water Permeability. *Advances in Seed Biology*, Editor: José Giménez-López ISBN: 978-953-51-3622-4 Print ISBN: 978-953-51-3621-7 doi: 10.5772/intechopen.68178 Chapter 10.

Garzón López, T. (2003). *Estudio de la compartimentación celular en plantas modelos sometidas a estrés por aluminio* [*Study of cell compartmentation in model plants subjected to aluminum stress*]. Tesis doctoral Facultad de Ciencias Universidad Autónoma de Barcelona. Barcelona, España. 95 pp.

Gunss, R. C. (1984). Fruits and seeds of genera in subfamily *Mimosoideae* (*Fabaceae*). *United States Department of Agriculture Technical Bulletin 1681*. Washington, DC, United States Department of Agriculture.

_________ (1991). Fruits and seeds of genera in subfamily *Mimosoideae* (*Fabaceae*). *United States Department of Agriculture Technical Bulletin 1755*. Washington, DC, United States Department of Agriculture.

Harper, L. W. (1991). The effect of harvesting and conditioning on seed dormancy. *Seed Technologist News*. 65, 2: 50-55.

Harrington, J. F. (1972). Seed Storage and Longevity. In: *Seed Biology* (ed. T. Koslowski), pp. 145-245. Academic Press, New York, USA.

Harris, W. H. (1987). Comparative ultrastructure of developing seed coats of "hard-seeded" and "soft-seeded" varieties of soybeans, *Glycine max* (L.) Merr. *Bot. Gaz.* 148 (3): 324-331.

ISTA (2003). International Seed Testing Association. *International Rules for Seed Testing*. Switzerland.

ISTA (2007). *International Rules for Seed Testing*, Edition 2007, International Seed Testing Association, Bassersdorf, Switzerland.

Jose, S.; Elinana, B. O.; Gomes, C. and Matsuokka, K. (1986). *Arq. Biol. Technol.* (Curitiba) 29: 633-650.

Kauss, H. (1985). Callose biosynthesis as calcium regulated process and possible relation to the induction of other metabolic changes. *J. of Cell Sci. Suplement* 2: 89-103.

Kirkby E. A and Römheld V. (2007). Micronutrients in plants physiology: functions, uptake and mobility. *Proceedings 543, The International Fertilizer* Society UK.

Marcos Filho, J. (2005). *Fisiología de sementes de plantas cultivadas* [*Physiology of seeds of cultivated plants*] (ed. Fundação de Estudos Agrários Luis de Queiroz), vol. 12, 495 pp. Piracicaba, SP, Brazil.

Martens, H., H. B. Jakobsen and O. B. Lyshede (1995). Development of the strofiole in seeds of white clover (*Trifolium repens* L.). *Seed Science Research*, 5, 171-176.

Mc Kee, G. W., R. A. Pfeiffer and N. N. Mosheninn (1977). Seed coat impermeability to water. *Agronomy Journal*, 69, 53-58.

Método 4500 – P C Vanadomolybdophosphoric Colorimetric Acid (1998). In: *Standard Methods for the Examination of Water and Wastewater*, 20th Edition, 1998, APHA, AWWA, WEF. Method 4500-P B and C.

Meyer, Ch. J., Steudle, E. and Peterson, C. A. (2007). Patterns and kinetics of water uptake by soybean seeds. *Journal Experimental Botany*, 58, 717-732.

Moravec, C. M., Bradford, K. J. and Laca, E. A. (2008). Water relations of drumstick tree seed (Moringa oleifera): imbibition, desiccation, and sorption isotherms. *Seed Science and Technology*, 36, 311-324.

Nikolaeva, M. G. (1980). Factors controlling the seed dormancy pattern. In: *The physiology and biochemistry of seed dormancy and germination.* (eds. Elsevier/North-Holland Biomedical Press), pp. 51-74, Amsterdam, The Netherlands.

Pammenter, N. W. and Berjak, P. (2000). Aspects of recalcitrant seed physiology. *Revista Brasileira de Sementes*, 12, 56-69.

Perissé, P. and Planchuelo, A. M. (2004). Seed coat morphology of *Lupinus albus* L. *and Lupinus angustifolius* L. in relation to water uptake. *Seed Science and Technology*, 32, 69-77.

Salisbury, F. B. and C. W. Ross (1992). *Fisiología de las plantas. Células: agua, soluciones y superficies* [*Plant physiology. Cells: water, solutions and surfaces*], Eds. Thomson, Paraninfo S.A., 305 pp., Madrid, España.

Sandoval, Z, E. (2005). Técnicas aplicadas al estudio de la anatomía vegetal [Techniques applied to the study of plant anatomy]. Universidad Nacional Autónoma de México. *Cuadernos*, 38, 51 pp., DF, México.

Simmons, G. F. and A. W. Riedl. (1968). *Hard seed of alfalfa*. Western Society of Crops Science Abstracts.

Taiz, L. and E. Zeiger. (1998). *Plant physiology*, 2ª Ed. Sinauer Associates, Inc. USA. 792 pp.

Vertucci, C. W. (1993). Predicting the optimum storage conditions for seeds using thermodynamic principles. *Journal of Seed Technol*. 17, 2, 41-53.

Yassen, M. Y., S. A. Barringer, W. E. Splittstoesser and S. Constanza (1994). The role of seed coat in seed viability. *The Botanical Review*, 60 (4), 426-439.

Zeng, L. W., P. S. Cocks, S. G. Kailis and J. Kuo (2005). Structure of the seed coat and its relationship to seed softening in Mediterranean annual legumes. *Seed Science and Technology*, 33, 351-362.

Zimmermann, L. R.; Galussi A. A.; Martinelli, A. HM.; Fernández, A. P.; García, A. H.; Pitter, J. R.; Zecchin, A. R. and D. I. Dechanzi. (1998). Hard seed viability in *Medicago sativa* L. *Lotus corniculatus* L.,

Trifolium repens L. *Trifolium pratense* L. y *Melilotus alba* Med. *Seed Science and Technology*, 26 (1) 271-273.

Zimmermann, L. R; Galussi, A. A; Martinelli, A. H. M. and Fernández, A. P. (2003). Viabilidad y longevidad de las semillas duras de leguminosas forrajeras [Viability and longevity of hard seeds of forage legumes].. *Ciencia Docencia y Tecnología*, Universidad Nacional de Entre Ríos, Argentina 26, 231-257

In: Plant Dormancy
Editor: Renato V. Botelho

ISBN: 978-1-53615-380-4

Chapter 10

Biochemical and Physiological Aspects of Dormancy in Temperate Fruits Trees

Silvia Scariotto[1], Américo Wagner Júnior[2] and Idemir Citadin[1,*]
[1]Department of Agronomy, Federal University of Technology - Paraná, Pato Branco, PR, Brazil
[2]Department of Agronomy, Federal University of Technology - Paraná Dois Vizinhos, PR, Brazil

Abstract

Dormancy is a resistance or adaptive mechanism that plants have developed as a way of surviving adversities during the winter period. Dormancy is not a rapidly acquired mechanism, but a progressive development process that begins during the fall, increasing in intensity until reaching the so-called deep dormancy or endodormancy during the winter. This phase can be induced by internal and external factors and is regulated by their interactions: environmental factors, such as photoperiod

* Corresponding Author's E-mail: idemir@utfpr.edu.br.

and temperature; and by physiological factors that include phytohormones and biochemicals (sugar, starch and carbohydrate content). The intensity and the way this phenomenon manifests in temperate plants, different growing environments, has an important impact on the control, maintenance and production of such species. Thus, a better understanding of the types of dormancy (paradormancy, endodormancy and ecodormancy) and the characterization of the growing environment are important to obtain satisfactory yields and to determine the most suitable agronomic techniques. This chapter presents to the reader the internal and external aspects related to the induction and exit of dormancy, as well as methods of evaluating the stage of dormancy.

Keywords: paradormancy, endodormancy, ecodormancy, environmental effect

1. Introduction

The growth and development of temperate fruits trees are related to changes during their production cycle, with the tress being conditioned to meteorological factors, mainly solar radiation, temperature and precipitation, with a specific response according to the species and/or cultivar. Temperature is responsible for the seasonal rhythm, which conditions the temperate fruit trees to present distinct periods of development, among them, (i) vegetative, which involve the full vegetative metabolic activity; (ii) reproductive: passing to the stage of induction and differentiation of new floral buds for the next cycle; and (iii) dormancy or hibernation, in which the trees do not show visible vegetative growth.

Although dormancy is a problem for many crops, it is very important for temperate fruit trees, as it allows the survival of plants under stress conditions brought about by low winter temperatures. When in the dormancy stage, buds need to go through an effective period of cold to stimulate uniform sprouting and flowering, that is, the cold ceases to be a stressful agent and becomes indispensable for the resumption of the new growth cycle of the plant. The depth of dormancy, and the need for cold accumulation to overcome it, is determined by the genotype x environment interaction and is a decisive factor in climatic zoning, allowing the recommendation or not of cultivars and/or fruit plants for cultivation within a certain area, as it interferes directly on the adaptation of the plant. Therefore, dormancy is a mechanism of resistance or adaption that plants have developed as a way to face climatic adversities, characteristic of the Northern Hemisphere (Bonhomme 1998).

Several authors have sought to conceptualize dormancy (Samish 1954; Vegis 1964; Saure 1985; Metivier 1985), but according to Camelatto (1990), none of them have defined it completely, which is perhaps due to the lack of specific knowledge regarding the aspects involved in the entry, maintenance and overcoming of dormancy. This occurs because the growth of the bud is conditioned by a combination of environmental factors (ecodormancy), the influence of another organ of the tree upon it (paradormancy) or is linked to specific biochemical and physiological events of the bud (endodormancy). These three terminologies paradormancy, endodormancy and ecodormancy define the phases of dormancy (Lang et al. 1985; Lang et al. 1987) (Table 1).

During the paradormancy phase, which can begin even during the summer after the fruit harvest, the onset of flower differentiation occurs (Fadón et al. 2015). In this phase, budburst is inhibited by other tissues through hormonal action and different types of competition between buds. At the end of this phase leaf senescence starts due to a combination of external factors, such as the occurrence of mild temperatures and photoperiod reduction, and is mediated internal by the balance of plant hormones and other signals that are not yet fully understood (Rogers 2017). The leaf nutrients are reallocated to parenchyma cells specialized in storage, to be used later in the initial support of sprouting and flowering during spring. The shortening of the photoperiod and colder weather signal a transition to the endodormancy phase (Crabbé and Barnola 1996; Campoy et al. 2012). In subtropical regions, the beginning of the endodormancy period can be considered when leaf fall ends.

During endodormancy, inhibition is established inside the bud itself, and buds do not develop even under favorable environmental conditions. This is crucial for plant survival at low temperatures. Although para- and ecodormancy are influenced by several factors, endodormancy is regulate predominately by low temperatures (Cook et al. 2017), because chilling accumulation removes the physiological blocks that inhibit budburst and flowering (Dennis 2003). In the ecodormancy phase, meristems achieve the full development capacity (Legave et al. 2008), but plants remain dormant if the environmental conditions are not favorable (Lang et al. 1987). This phase is overcome by a certain amount of accumulated warmth. The transition between endodormancy and ecodormancy phases is difficult to predict. The development of floral buds in mild winter regions are characterized by the abrupt resumption of development after endodormancy release, with a very short or non-existent ecodormancy (Malagi et al. 2015). The cultivation of temperate fruit trees in mild winter regions and the current climatic changes have led to a resumption of the study of dormancy, its causes and consequences. Understanding the dormancy phenomenon and its relation to environmental factors, under special conditions of insufficient chilling, is crucial for the development of cultivars that are better adapted to the future scenarios of global warming. An increase in temperature during late fall and winter in several regions would reduce the chilling accumulation (Baldocchi e Wong 2008; Luedeling et al. 2011).

Table 1. Stages of dormancy and the factors that induce each of them

Sprouting stages		
Ecodormancy	Paradormancy	Endodormancy
Regulated by environmental factors	Regulated by external physiological factors to the affected structure	Regulated by internal physiological factors to the affected structure
Examples		
Extreme temperature Nutrient deficiency Hydric stress	Apical dominance Influence of photoperiod	Response to the cold Influence of photoperiod

Source: Lang et al. (1987).

In California specifically, by the year 2100; the accumulation of chilling hours will be reduced by 40 hours per decade and, by the end of the 21st century, the total accumulated will be only 500 chilling hours per winter, which will have a great impact on local fruit growing (Baldocchi e Wong 2008).

The aim of this chapter is to present the reader with a review of dormancy, its biochemical and physiological aspects, as well as to address external factors (temperature and photoperiod) related to the induction and overcoming of dormancy in buds of temperate species, especially when grown in mild winter regions.

2. Bud Dormancy

After the perception of the decrease in photoperiod, during the autumn, by the leaves of temperate fruit trees, there are several hormonal stimulus that induce the synthesis of enzymes that will degrade the middle lamella and the cell wall of the petiole, causing the fall of the leaves and inducing the plant to enter into dormancy. Although it does not show visible growth during this phase, metabolic activities continue, but with reduced intensity, allowing them to withstand low temperatures, even lower than 0 °C (Petri et al. 1996). In humid subtropical regions, like the temperate fruit production zone in Brazil, defoliation is often caused by foliar diseases, such as rust (*Tranzschelia discolor*) and bacterial spot (*Xanthomonas arboricola* pv. *pruni*) in peach trees, 'glomerella' leaf spot (*Colletotrichum gloeosporioides*) and scab (*Venturia inaequalis*) in apple. Early defoliation alters the dynamics of dormancy, reduces the storage of carbohydrates and hasten flowering and sprouting.

From the various definitions given for plant dormancy, Saure (1985) described it as a resting phenomenon of deciduous plants, termed pre-dormancy, true dormancy and imposed dormancy. Pre-dormancy is the stage where sprouting of the lateral buds is inhibited directly by the growth of the apex of the branches and/or the adjacent leaves. In true dormancy, the growth of the buds is inhibited by the endogenous system, located in the bud itself, which is characteristic of each species or cultivar and of each bud in the plant. True dormancy is divided into three periods, initial, deep, and final. With the end of true dormancy inhibition of the growth of buds by exogenous conditions can occur, directly and reversibly imposed by the environment, which is called imposed dormancy.

In spite of the various definitions of dormancy, in general, the most used and accepted is the temporary stop of the growth and visible development of any structure of the plant that contains meristem, which, in turn, can be divided into para-dormancy, eco-dormancy and endo-dormancy (Lang 1987). In general, the prefixes eco-, para- and endo- indicate that inhibition of growth can move from outside the plant (ecodormancy) to inside the plant, but out of the bud (paradormancy) and into the bud itself (endodormancy) (Cook et al. 2017).

Paradormancy is characterized by correlative inhibitions that are controlled by physiological factors external to the structure in which dormancy manifests. In inhibition caused by apical dominance, for example, apical buds interfere with basal bud growth, and it is possible they immediately restart their growth by suppression of the inhibitor. Both young and adult leaves may also be a source of inhibition or competition for water and nutrients (Crabbe 1970). According to Lauri (2007), the influence of apical dominance was coordinate by basipetal transport of auxin, which exerts direct inhibition on the growth of basal buds. Auxin also exhibits signaling functions, in part, by specific protein degradation and regulation of cytokinin production (auxin inhibits or decreases cytokinin sensitivity) in the adjacent stem segments of the basal buds (Shimizu-Sato and Mori 2001 Stirnberg, Van De Sande, and Ottoline 2002). Cytokinin is responsible for inducing the expression of D-type cycling (CYCD3) and cyclin-dependent B-type kinases (CDKBs) necessary for cell division.

In the endodormancy, the growth is inhibited due to physiological changes in the bud and it is regulated by chilling. It is accepted that the endodormancy period is divide into two parts, d-endodormancy (deep) and s-endodormancy (shallow) (Faust et al. 1997). The first occurs after the paradormancy, with a possible overlap, until the endodormancy peak, with the bud unable to sprout, even if the environmental conditions are favorable. In this phase the cumulative effect of the chilling is essential to eliminate inhibitors and to stimulate the development of the bud, and the chemical agents for the break of dormancy do not take effect in substitution of the chilling. The S-endodormancy is the period after the peak of endo-dormancy, when the use of chemical agents can replace chilling, hastening and homogenizing the restart of growth (Faust et al. 1997) (Table 3).

The phenomenon of endodormancy is induced more slowly in lower latitudes than in typical temperate zones (Erez 2000). However, in humid subtropical regions, in years of low chilling accumulation, the plants do not enter into deep endodormancy (Malagi et al. 2015; Carvalho and Zanette 2004a-b), remaining in superficial endodormancy throughout the rest period. In productive systems, this phase is the most worrisome for producers, since poor or ununiform sprouting and blooming may decrease fruit production and future development of the plant.

Table 3. Factors that determine the entry and exit of dormancy in temperate fruit trees

Budding State		
Paradormancy	d-endodormancy	s-endodormancy
Inducing Factors		
Exogenous		
Short days (SD) in autumn, indoleacetic acid (IAA), acid abscisic, terminal buds, leaves	Low temperatures in late autumn	Ending of cold accumulation
Endogenous		
	Dehydrins, changes in membranes, water fixation, limited metabolic activity	Interactions AIA / cytokinin, free water, metabolic activity
Output Factors		
Defoliation, Thidiazuron (TDZ), CN_2H_2, KNO_3, elimination of shoots, water stress + irrigation	None	Elevating temperature, TDZ, CN_2H_2, KNO_3,

Source: Faust et al. (1997).

The entry into endodormancy is influenced by the conditions of the photoperiod and decreasing temperatures, which induce the leaves to produce growth inhibitors, mainly abscisic acid (ABA).

The accumulation of this causes the reduction of branch growth, as a consequence the internodes will be shorter. At the end of the leaves abscission, the growth of the branches and the production of the inhibitors completely cease. The ratio between promoters (example: gibberellins and cytokinin) and inhibitors are low during endodormancy. In the winter, with the cold action, the gradual reduction of the growth inhibitors and an increase in the synthesis of growth promoters starts. This phenomenon occurs until the endodormancy is overcame and the plant initiates a new cycle of growth. Thus, only after a certain amount of chilling hours are accumulated, the bud will begin to grow if the spring temperature is favorable (Cook et al. 2017).

The scales of the buds have a significant participation in the endodormancy phenomenon, since they limit the diffusion of oxygen in the meristems, which causes carbon dioxide accumulation to levels that reduces the respiratory activity of these tissues. In addition, the scales act at a physiological level, as their tissues contain a large amount of growth inhibitors with a possible action on the metabolic activity of the meristem (Leopold and Kriedemann 1975).

The transition from endodormancy to ecodormancy is marked by the hydration capacity of the buds and it occurs gradually in temperate regions, because normally the temperature is not yet sufficient to activate the metabolism rapidly (Cronjé et al. 2003; Malagi et al. 2015).

On the other hand, under mild weather, the transition between endodormancy and ecodormancy is very fast (less than 10 days), since it occurs at a temperature generally above 15 °C, which hastens the plant metabolism (Malagi et al. 2015). Even though endodormancy has not been completely overcome, exposure to high temperatures induces sprouting (Tromp et al. 2005). However, the time spent sprouting varies according to the state of dormancy of the bud.

In general, several hypotheses were haven been formulated regarding the factors involved in the entry, maintenance and exit of endodormancy.

Therefore, dormancy may be related to energetic metabolism changes in the bud, such as enzyme activity and nucleotide synthesis (Bonhomme et al. 2000); hormonal regulation (Stafstrom 2000); carbohydrate content and flow, nutrient supply and translocation of short-range reserves (Carvalho 2001; Erez 2000; Marquat et al. 1999); and protein content (Tamura et al. 1998).

During the summer/autumn with photosynthetic production, the storage of starch and carbohydrate in the cells of the parenchyma of the xylem of the branches of perennial plants is necessary to avoid frost damage during winter and to provide energy in resumption of active growth (Sauter 1988; Borba, Scarpare Filho, and Kluge 2005). The stored starch is converted to soluble sugar during endodormancy in response to low temperatures, which helps to prevent freezing and/or to recover vessel cavitation in plants exposed to freezing temperatures (Sauter et al. 1996; Améglio et al. 2001).

According to the need of the plant, the starch can be hydrolyzed into simple sugars that can be used in three ways: in respiration, for the formation of adenosine triphosphate (ATP) and other important compounds; stored in the vacuole for later use; and resynthesized again into starch to be stored in the drain tissue, serving as a reserve for use in situations of lower photosynthetic activity (Borba, Scarpare Filho, and Kluge 2005).

The profile of the carbohydrates present in the plant can change during endodormancy, which consequently influences the activity of enzymes such as alpha-amylase and sucrose phosphate synthase (SPS), responsible for the hydrolysis of starch and sucrose synthesis, respectively. Studies related to the dynamics of carbohydrates in plants encompass the trophic theory of dormancy (Alves et al. 2007; Decourteix et al. 2008; Citadin et al. 2009; Bonhomme et al. 2010). This theory relates to the ability of plants to sprout with the ability of the buds to acquire and utilize carbohydrates (Tort et al. 1985, Gendraud and Pételt 1990).

Among carbohydrates, sorbitol is the most studied because it is one of the main translocate carbohydrates of the species in the Rosaceae family (Moing et al. 1997). Studies have shown that sorbitol content and the ability of the tree to use them coincide with some important points of the dormancy period (Maurel et al. 2004; Ito et al. 2012; Ito et al. 2013). In peach trees, a decreasing concentration of sucrose and sorbitol occurs during the winter and an increase in glucose and fructose just prior to sprouting (Maurel et al. 2004). In walnut trees, an increase in bud sucrose is only observed after endodormancy (Bonhomme et al. 2010). According to Ito et al. (2012), the Japanese pear modulates the catabolism of sorbitol and sucrose according to the stage of the dormancy buds. In addition, they suggest an inhibitory effect of sorbitol on the growth of floral buds during the period of endodormancy. The functional role of sorbitol is different to that of sucrose during the endodormancy period. Sorbitol has a protective function against abiotic stress caused by low temperatures (Kanayama 2009).

In the period before sprouting and flowering, mainly in the apical portion of the branches, α-amylase activity (Citadin et al. 2009), and the hexoses (Bonhomme et al. 2010) increase in walnut trees. At this stage, sucrose reduction also occurs due to the high activity of the SuSy enzymes and/or invertases enzymes involved in sucrose catabolism, which are considered, in the literature, as good markers of sink strength (Morris and Arthur 1984; Roitsch and Gonzalez 2004). There is also an increase in ATPase activity and acid invertase (IA), responsible for the entry of soluble sugars into the cells of the sink tissues (Alves et al. 2007). Bonhomme et al. (2005) observed that the vegetative primordia, especially the floral, presented high concentrations of transportable forms of carbohydrates (sucrose and sorbitol), when the growth capacity is recovered at the endodormancy output. Bonhomme et al. (2010) observed a correlation between the influx of sugars (mainly sucrose) and bud growth rates.

The intensity of the dormancy of the buds is directly related to the catalase activity, which shows a marked reduction with temperature decline in winter. The decrease in catalase activity causes an increase in the hydrogen peroxide levels in the tissues of the buds, which activates the phosphate-pentose metabolic pathway and the sprouting, followed by a fast growth (Nir and Shulman 1986).

During the last stages of endodormancy and during ecodormancy, the buds are sensitive to cytokinin and other chemical substances that induce the overcoming of dormancy by exposing them to high temperatures (Agusti 2004). However, the application of chemical products to stimulate the break in dormancy only takes effect when part of the chilling requirement has been satisfied (Donoho and Walker 1957, Weinberger 1969, Lavee 1973). There is not chemical product that completely replaces the effect of chilling, only partially. Such products have greater importance in places where trees do not enter into deep endodormancy or have already overcome it.

Although auxins are necessary to activate bud growth, they have limited involvement in the control of bud dormancy (Lavee and May 1997). Since the 1950s, it has been shown that endogenous auxin levels in buds decrease considerably with increasing endodormancy, but with swelling before sprouting the levels increase again (Blommaert 1954; Hendershott 1959). Thus, it can be concluded that auxins are not involved in the control of the overcoming of endodormancy, but rather in the development of the bud (Lavee and May, 1997).

When the buds are close to sprouting the synthesis of the enzyme precursor of ethylene (the 1-aminocyclopropane 1-carboxylic acid - ACC) is induced, and once exposed to continuous chilling, it converts ACC to ethylene (Putti 2001). After this, the ethylene will increase the permeability of the cell membrane, allowing the sprouting of the vegetative buds. However, ethylene is not effective at breaking dormancy in apple tree buds (Paiva and Robitaille 1978) and grapevines (Lavee and May 1997), but application of ethephon (2-chloroethylphosphoric acid) causes delay in sprouting (Szyjewicz, Rosner and Kliewer 1984).

Proteins also exhibit variations in content during the endodormancy period. Studies show that the regulation of the gene responsible for the production of dehydrin protein is linked to the cold acclimation in peach tree tissues and the xylem, increasing in concentration during this period (Arora and Wisniewski 1994, 1996; Artlip et al. 1997; Wisniewski, Bassett and Arora 2004). Dehydrins are also associate with cold acclimation in blueberry (Muthalif and Rowland 1994) and pear tree (Tamura et al. 1998). Three dehydrins are strongly induced by cold in the blueberry flower shoots, and the levels of these are correlated with the levels of resistance to cold (Muthalif and Rowland 1994). In apple tree, transcripts of MdDHN, a dehydrin isolated from this species, are highly expressed in bark tissue and buds when it the tree is in endodormancy, but it is not express in early spring when cold resistance is lost and the shoots are growing (Garcia -Bañuelos et al. 2009). It is proposed that dehydrins can protect proteins and membranes from unfavorable structural changes caused by dehydration. In vitro experiments have revealed that the presence of this protein promotes the maintenance of the catalytic activity of α-amylase and the appropriate level of starch hydrolysis during the decrease in temperature. This suggests that dehydrins maintain the adequate concentration of water near α-amylase, which is an important precondition for enzyme-substrate complex formation and starch degradation (Rinne et al. 1999). The water retention capacity of the dehydrins is also supported by peach tree studies. The helical sites of these proteins determine their affinity for the surface of the ice crystal network, and this prevents the subsequent growth of ice crystals (Wisniewski et al. 1999).

Fatty acids also play an important role in the development and break of endodormancy, since their role has been attributed to changes in cell membranes during each phase of this process (Wang and Faust 1990).

3. Methods for Estimating Endodormancy

The precise assessment of endodormancy in trees under field conditions is difficult, since solar radiation, daily temperature and other factors cannot be controlled. Most studies have been conducted under controlled environments, mainly for temperature. There are a lot of tests to evaluate dormancy, both qualitative and/or quantitative, which observe the state and/or depth of dormancy of the buds (Mauget 1987). The use of tests with more practical applications allows the stud of the transition between the phases of dormancy, their duration and the variation between different climatic conditions.

Among the several tests used to estimate endodormancy of a species and/or a variety there are the detached branch test (Citadin et al. 1998; Citadin 1999; Herter et al. 2000; Citadin et al. 2001; Wagner Júnior et al. 2009); biological method or single-node cutting test (Pouget 1963; Rageau 1978; Dennis 2003; Herter et al. 1992; Balandier et al. 1992; Balandier et al.

1993; Bianchi et al. 2000); potted plants test (Camelatto et al. 2000; Citadin et al. 2001; Silveira 2003); and the Tabuenca's test (Tabuenca, 1964; Bonhomme et al. 1997; Malagi et al. 2015).

The methodology of grafted shoots has shown promise in the evaluation of the effect of artificial chilling hours on peach tree dormancy (Silveira, 2003; Wagner Júnior et al. 2009). One of the advantages of this method is the greater longevity of the branches during the evaluation, after it was grafted. However, this method is more labor-intensive and requires more space in the growth chamber and greenhouse than in the detached branches test, which may limit its use depending on working conditions.

The single-node cutting test allows evaluation of the endodormancy depth of the buds, from different portions of the branch, since the buds are isolated, that is, the effect of the paradormancy is eliminated. For this test, the branches are broken into parts containing a single bud, then submitted to the temperature and photoperiod suitable for growth (Rageau 1978; Balandier et al. 1993; Bonhomme et al. 2000; Carvalho et al. 2010; Carvalho and Biasi, 2012). The limitations of this method are that it is not possible to evaluate the growth and development of shoots; and the sprouting percentages obtained are often higher than those obtained from other methodologies, especially for axillary buds, which is justified by the elimination of most of the correlative inhibitions (Champagnat 1983; Mauget and Rageau 1988; Balandier et al. 1992).

Tabuenca's test is used in some studies to determine the moment of the exit of endodormancy, by means of the evaluation of the increase in fresh matter and/or dry matter of the floral primordia (Bonhomme et al. 1997; Leite et al. 2006; Malagi et al. 2015).

According to Saure (1985), the 50% sprouting rate in temperate species can be used as an indicator of the break in endodormancy. However, the innumerable characteristics that determine sprouting in this test limit the use of the results for the physiological study of overcoming dormancy and endodormancy (Dennis 2003). The flowering percentage is another parameter used as an indicator for the break in endodormancy. In apple trees, flowering above 80% is considered the overcoming of endodormancy

(Shaltout and Unrath 1983). In peach trees, the adopted criteria is 50% of flowers open (Richardson et al. 1974).

Biochemical tests (nucleotide tests) have been developed based on the ability of adenosine triphosphate (ATP) synthesis by vegetative buds (Bonhomme et al. 2000). There are also those that are based on the determination of the carbohydrate content of the buds and branches to detect the deviation of the photoassimilate flow to the tissues near or adjacent to the meristematic zone, which allows the elaboration of curves that show their dynamics (Carvalho and Zanette 2005).

Other studies to evaluate endodormancy consider variations in the fatty acid content of bud cell membrane phospholipids (Erez 2000), the amount and activity of ATPase in the plasma membrane of adjacent buds and tissues (Aue et al. 1998; Aue et al. 2000), the respiration intensity of the buds during autumn and winter (Mcpherson et al. 1997), and in pollen meiosis (Citadin et al. 2002).

4. Genetic Endo-Dormancy Approach

The expression of several genes involved in endodormancy in deciduous plant buds has been characterized. Studies on apple trees have shown that the expression of KNAP2 (I KN1-like gene) increases during the start of endodormancy and decreases with the end of endodormancy (Brunel et al. 2002).

A peach mutant tree, known as "evergrowing" (EVG) identified in southern Mexico does not stop growth, but acquires endodormancy under induction conditions (Rodriguez et al. 1994). In the analysis, six MADS-boxes (dormancy associated MADS-box), transcripts were identified but not expressed in the EVG mutant (Bielenberg et al. 2008). Among the six genes associated with dormancy (DAM), DAM1, DAM2 and DAM4 present expression patterns related to the suspension of growth and bud formation, are the candidates most likely to control the EVG phenotype. DAM3 does not respond to photoperiod and it could be regulated by low temperatures. DAM5 and DAM6 showed an increase during the fall, with a peak coinciding with the winter solstice, followed by a rapid decline (Li et al. 2009).

Based on the published results and those obtained by Jiménez et al. (2010), hypotheses are proposed for the role of DAM5/DAM6 in establishing the chilling requirement of buds and in regulating the transition from endodormancy to ecodormancy. Short photoperiods trigger the expression of DAM5/DAM6 in the fall. Low temperatures indirectly reduce DAM5/DAM6 expression by interrupting the perception of photoperiodic stimuli and/or by a direct effect on the transcription of DAM5/DAM6. DAM5/DAM6 directly represses or promotes the expression of genes that control bud tissue growth under appropriate conditions. Zhong et al. (2013) observed that the genes DAM3, DAM5 and DAM6 were abundantly expressed during endodormancy, decreasing their expression during dormancy. As DAM5 and DAM6 were negatively regulated by prolonged exposure to cold in both, peach and apricot trees (Jiménez et al. 2010; Yamane et al. 2011a, b; Sasaki et al. 2011), these genes may contribute to the overcoming of endodormancy in *Prunus*.

Fan et al. (2009) found a strong quantitative trait loci (QTL) linked to chilling requirement (CR) and blossom time co-located with the locus EVG. The co-localization of the EVG locus with the QTL associated with the CR characteristic suggests that these DAM genes are potential candidate for endodormancy. Peach tree DAM genes are members of the SVP/StMADS11 lineage (Jiménez et al. 2009).

Yamane et al. (2008) identified a MADS-box gene associated with endodormancy. This gene had a high similarity with the Short Vegetative Phase (SVP)/Agamous-Like24 (AGL24), which is regulated positively during onset and negatively during the overcoming of endodormancy in apricot tree lateral buds (*Prunus mume*). SVP inhibits flowering through the down-regulation of a floral integrator FLOWING THE LOCUS T (FT) (Hartmann et al. 2000; Lee et al. 2007), while AGL24 acts as a promoter of flowering (Michaels et al. 2003). Differential expression has also been observed during dormancy for hormone-linked genes, such as ABA, gibberellic acid (GA) and ethylene (Rohde et al. 2002; Rohde and Bhalerao 2007). Zhong et al. (2013) compared the expression of genes responsible for the production of auxins, ABA and ethylene during transitions from one dormancy stage to the next in *Prunus mume*, observing that some genes involved in auxin responses showed a higher degree of expression during the transition from paradormancy to endodormancy. This suggests that auxin-related genes may play an inhibitory role in the transition from paradormancy to endodormancy in apricots trees. There are lower levels of expression of genes linked to ethylene ERF1 (ppa009707m) and ERF5 (ppa007892m) in endodormancy than in paradormancy. In contrast, the ABA-related genes (ppa006696m and ppa008716m) are regulated in the endodormancy stage compared to the paradormancy stage. Results obtained suggest that ethylene and ABA can play an interactive role in buds dormancy.

5. Endodormancy Environmental Effect

Photoperiod and temperature play a significant role in the induction and overcoming of endodormancy. The shortening of day length induces a change in development in the terminal shoots that results in the formation of scales rather than leaf buds (Okuba 2000). Tanino et al. (2010) proposed the existence of two separate, but temporally connected, processes that contribute to the development of endodormancy in some woody plants: one directed by the photoperiod and influenced by moderate temperatures; the other directed by abiotic stresses, such as low temperature in combination with long photoperiods. The molecular changes corresponding to these two related, but distinct, responses remain an investigative challenge.

Baba et al. (2011) observed that the short days influenced the cellular division of the cambium, causing its stoppage. This may have occurred due to the alteration in the auxin response that indirectly leads to the low regulation of genes responsible for the cell cycle, thus inducing growth stoppage. This alteration can also negatively affect the hormonal pathways of ethylene and gibberellin synthesis that promote the division of the cambial cells and also the transition from the growth phase to the endodormancy phase (Eriksson et al. 2000; Ruonala et al. 2006).

The occurrence of cold and its effect on the onset and exit of endodormancy in temperate fruit trees was evaluated in terms of duration (quantitative aspect) and intensity (qualitative aspect) (Silveira 2003). The effects for exit endodormancy are not evenly distributed in the tree, that is, when part of the tree is under conditions to accumulate cold while others are not, the sprouting of the buds will occur only in those parts exposed to the cold (Salisbury and Ross 1992).

When the cold requirement is not satisfied, temperate fruit species present several physiological disorders, the severity of which are dependent on the extent of the cold deficit to which they were subjected. The main symptoms are a reduction in sprouting and/or flowering, lack of uniformity of budburst (temporal or spatial erraticism), with few leaves, resulting in problems in tree formation, productivity and fruit quality, besides the promotion of changes in the tree cycle. In addition, other abnormalities such as, pistil abortion, alteration in pollen growth, leaf deformation, appearance of multiple pistils leading to multiple fruits, necrosis of buds that fall before budbreak, and flowering (Petri and Leite 2004; Oukabli and Mahhou 2007). Normally, when this happens the development cycle is altered, with the occurrence of a greater period of vegetative growth and a reduced period without leaves, besides compromising the quality of the production. Thus, the knowledge of the chilling accumulation of a given region and the chilling necessity of each variety and/or species is important, because in regions where the tree can acquire their chilling needs, it is possible for them to express their maximum agronomic potential.

In a study on Japanese pear trees submitted to cold deprivation, Marafon et al. (2011) noted that: not induce the mobilization of the starch; the synthesis of sucrose in the woody tissues (source) and its export to the floral buds (sink) were reduced; the concentrations of total and reducing sugars, activities of acid invertase enzymes and sucrose phosphate synthase were lower in branch tissues with chilling deprivation; while the starch concentration was higher in the wood of plants that did not receive chilling. This reinforces the hypothesis that the erraticism observed in the budburst and flowering of temperate climate fruit trees that are not adapted, is a consequence of the irregularity in the mobilization of carbohydrates.

The hydrolysis of starch in stems and regions adjacent to the buds in trees maintained at 15 °C during endodormancy occurs later than in trees maintained under ideal chilling conditions. This is support by the fact that: under deprivation of chilling the hydrolysis of the starch is not associated with an increase in hexose in stems and regions adjacent to the buds; and that in natural cold conditions the metabolic activity is related to the biosynthesis of soluble sugars (Bonhomme et al. 2005).

Plants with chilling deprivation showed significant differences in carbohydrate concentration among the different tissues, such as, the higher concentration of the carbohydrate in the adjacent regions and in the scales of the vegetative and floral buds, in relation to the primordia. The deviation of carbohydrates to other tissues (regions adjacent to the buds and scales) may be a consequence of the low sink strength exerted by the vegetative and floral primordias.

Sucrose and sorbitol concentrations remain high in the early stages, however, due to the lack of growth and the inability to use them by the primordia, cause necrosis (Bonhomme et al. 2005). The occurrence of abortion, delay in budbreak and necrosis of floral primordium are common in temperate fruit trees when grown in tropical or subtropical conditions, when chilling accumulation is insufficient or when cold occurs with episodes of moderate temperature (>15°C) (Weinberger 1967; Legave 1978; Legave and Garcia, 1982; Balandier et al. 1992, Viti e Monteleone 1993).

Vegetative buds have a greater chilling requirement than reproductive ones, consequently the blooming occurred before the sprouting, as the lateral buds of the shoots have a greater chilling requirement than the terminal buds, with that, the budbreak of the terminal buds occurs before that of the lateral buds (Petri et al. 1996). However, in years with a prolonged chilling period, above the real requirement of the peach varieties, there may be sprouting ahead of blooming (Citadin et al. 2001, Citadin et al. 2003).

After the chilling requirement for the break of endodormancy, it is necessary that a certain heat accumulation occurs, to activate the metabolism that leads to the new cycle of growth that begins with sprouting and flowering. The amount of heat requirement to stimulate metabolic processes varies according to species and/or variety (Citadin et al. 2001). However, the amount of heat requirement during ecodormancy is influenced by the amount of chilling accumulated during the winter (Okie et al. 2011). During mild winter years, the need for heat is greater than in winter years with greater chilling accumulation, that is, the greater the exposure to chilling in endodormancy the lower the heat requirement in ecodormancy. In general, the higher the temperature in this period, the faster flower buds grow (Citadin 1999). Thus, in temperate regions fruit species naturally have a greater ecodormancy than in subtropical regions (Malagi et al. 2015).

The effect of precipitation on endodormancy is not well understood, but there is a hypothesis that the cooling of buds through the use of intermittent sprinklers during the winter season is effective at stimulating budbreak, since it reduces and maintains the lower temperature of the buds during the endodormancy, thus avoiding the negative effects caused by the elevation of temperature. In apple and pear trees, winter rains reduce the period of endodormancy (Petri et al. 1996).

Thus, in regions or years where colder and prolonged chilling occurs there is the induction of deeper endodormancy and slower resumption of active growth. On the contrary, mild winter induces superficial endodormancy and abrupt rehydration of buds and a rapid resumption of development under heating (short ecodormancy), but often with erraticism of budburst and flowering. Understanding the dynamics of dormancy in contrasting conditions of chilling accumulation, especially in mild winter regions, is fundamental for the development of more precise strategies for adaptation and management of temperate fruit species under climatic change scenarios with climatic warming. Thus, the phenological monitoring for several years using standard species and varieties at contrasting chilling accumulation sites is suggested. It is also necessary to standardize methodology for the study of dormancy, with the Tabuenca Test (1964, 1967) and the single-node cutting test (Rageau, 1978) recommend as an end of endodormancy indicator and to analyze the dynamics and depth of dormancy respectively.

References

Alves, G., Decourteix, M., Fleurat-Lessard, P., Sakr, Soulaiman, B. M., Améglio, T., Lacointe, A., Julien, J. L., Petel, G., and Guilliot, A. (2007). Spatial activity and expression of plasma membrane H+-ATPase in xylem of walnut tree (*Juglans regia* L.), during dormancy and growth resumption. *Tree Physiology*, 22: 1471-1480.

Améglio, T., Ewers, F. W., Cochard, H., Martignac, M., Vandame, M., Bodet, C., and Cruiziat, P. (2001). Winter stem xylem pressure in walnut trees: effects of carbohydrates, cooling and freezing. *Tree Physiology*, 21:387-394.

Arora, R., and Wisniewski, M. (1994). Cold acclimation in genetically related deciduous and evergreen peach. II. A 60-kilodalton bark protein in cold-acclimated tissues of peach is heat stable and related to the dehydrin family of proteins. *Plant Physiology*, 105: 95-101.

Arora, R., and Wisniewski, M. (1996). Accumulation of a 60-kD dehydrin protein in peach xylem tissues and its relationship to cold acclimation. *Hort Science*, 31:923–925.

Arora, R., and Wisniewski, M. (2000). Seasonally-regulated proteins in peach (Prunus persica): what are they and what do they do? In *Dormancy in plants: from whole plant behaviour to cellular control*, edited by Viémont, Jean-Daniel, and Crabbé, Jacques, 161–72. Wallingford: CAB International.

Artlip, T. S., Callahan, A., Bassett, C. L., and Wisniewski, M. (1997). Seasonal expression of a dehydrin gene in sibling deciduous and evergreen genotypes of peach. *Plant Molecular Biology*, 33: 61-70.

Aue, H. L., Lecomte, I., Gendraud, M., and Pétel, G. (1998). Activity and quantitative analysis of plasmalemma ATPase in buds and bud stands during winter vegetative bud's dormancy in peach-tree. *Acta Horticulturae*, 465: 531-537.

Aue, H. L., Lecomte, I., and Pétel, G. (2000). Changes in the parameters of the plasmalemma ATPase during peach vegetative bud dormancy. *Biologia Plantarum*, 43: 25-29.

Baba, K., Karlberg, A., Schmidt, J., Schrader, J., Hvidsten, T. R., Bako, L., and Bhaleraoa, R. P. (2011). Activity–dormancy transition in the cambial meristem involves stage-specific modulation of auxin response in hybrid aspen. *Proceedings of the National Academy of Sciences of the United States of America*, 108: 3418-3423.

Balandier, P., Capitan, F., Quignot, S., Rageau, R., et Parisot, E. (1992). Etude de la croissance et du développement des bourgeons du pêcher cultivé à l'île de la Réunion [Study of the growth and development of the peach buds grown on the island of Reunion. Application to the development of a cultural itinerary]. Application à l'élaboration d'un itinéraire cultural. *Fruits, The International Journal of Tropical and Subtropical Horticulturae*, 47: 679-689.

Balandier, P., Bonhomme, M., Rageau, R., Capitan, F., and Parisot, E. (1993). Leaf bud endodormancy release in peach trees: evaluation of temperature models in temperate and tropical climates. *Agricultural and Forest Meteorology*, 67: 95-113.

Baldocchi, D., and Wong, S. (2008). Accumulated winter chill is decreasing in the fruit growing regions of Califórnia. *Climatic Change*, 87 (I): S153-S166.

Bianchi, V. J., Arruda, J. J. P., Casagrande Júnior, J. G., e Herter, F. G. (2000). Estudo da paradormência em pereira por meio do método biológico [Study of paradormancy in pear tree by means of the biological method]. *Revista Brasileira de Fruticultura*, 22: 294-296.

Bielenberg, D. G., Wang, Y., Li, Zhigang, Z., Tetyana, F., Shenghua, R., Gregory L., Scorza, R., and Abbott A. G. (2008). Sequencing and annotation of the evergrowing locus in peach *Prunus persica* (L.) Batsch reveals a cluster of six MADS-box transcription factors as candidate genes for regulation of terminal bud formation. *Tree Genet Genomes*, 4: 495-507.

Blommaert, K. L. J. (1954). The significance of auxin and growth substances in relation to winter dormancy of the peach tree. *Science Bulletin of the Department of Agriculture of South Africa*, 368: 45-53.

Bonhomme, M., Rageau, R., Richard, J. P., and Gendraud, M. (1997). Dormancy of peach floral buds: Biological and tentative biochemical approaches. *Acta Horticulturae*, 441: 167-74.

Bonhomme, M. (1998). *Physiologie des bourgeons végétatifs et floraux de pêcher dans deux situations thermiques contrastées pendant la dormance: capacite de croissance, force de puits et répartition des glucides* [*Physiology of vegetative and floral peach blossoms in two contrasting thermal situations during dormancy: growth capacity, well strength and distribution of carbohydrates*]. PhD diss., Université Blaise Pascal.

Bonhomme, M., Rageau, R, and Gendraud, M. (2000). ATP, ADP and NTP contents in vegetative and floral peach buds during winter: are they useful for characterizing the type of dormancy? In *Dormancy in plants: from whole plant behaviour to cellular control*, edited by Viémont, J. D., Crabbé, J., 245–57. Wallingford: CAB International.

Bonhomme, M., Rageau, R., Lacointe, A., Gendraud, M. (2005) Influences of cold deprivation during dormancy on carbohydrate contents of vegetative and floral primordia and nearby structures of peach buds (*Prunus persica* L. Batch). *Scientia Horticulturae*, 105: 223–240.

Bonhomme, M., Peuch, M., Ameglio, T., Rageau, R., Guilliot, A., Decourteix, M., Alves, G., Sakr, S., and Lacointe, A. (2010). Carbohydrate uptake from xylem vessels and its distribution among stem tissues and buds in walnut (*Juglans regia* L.). *Tree Physiology*, 30: 89-102.

Borba, M. R. C., Scarpare Filho, J. A., and Kluge, R. A. (2005). Levels of carbohydrates in peaches submitted to different intensity of green pruning in tropical climate. *Revista Brasileira de Fruticultura*, 27: 68-72.

Brunel, N., Leduc, N., Poupard, P., Simoneau, P., Mauget, J. C., Viemont, J.D. (2002). KNAP2, a class I KN1-like gene is a negative marker of bud growth potential in apple trees (*Malus domestica* [L.] Borkh.). *Journal of Experimental Botany*, 53: 2143-2149.

Camelatto, D. (1990). Dormência em fruteiras de clima temperado [Dormancy in temperate fruit trees]. *Horti Sul*, 1: 12–17.

Camelatto, D., Nachtigall, G. R., Arruda, J. J. P., e Herter, F. G. (2000). Efeitos de flutuações de temperatura, horas de frio hibernal e reguladores de crescimento no abortamento de gemas florais de pereira [Effects of temperature fluctuations, hours of chilling and growth regulators on the abortion of pear flower buds]. *Revista Brasileira de Fruticultura*, 22: 111-117.

Campoy, J. A., Ruiz, D., Allderman, L., Cook, N., and Egea, J. (2012). The fulfilment of chilling requirements and the adaptation of apricot (Prunus armeniaca L.) in warm climates: an approach in Murcia (Spain) and the Western Cape (South Africa). *European Journal of Agronomy*, 37: 43–55.

Carvalho, R. I. N. (2001). *Dinâmica da dormência e do conteúdo de carboidratos e proteínas em gemas vegetativas e ramos de um e dois anos de macieira com ou sem frio suplementar* [*Dynamics of dormancy and the content of carbohydrates and proteins in vegetative buds and branches of one and two years of apple tree with or without supplementary chilling*]. PhD diss., Universidade Federal do Paraná.

Carvalho, R. I. N., and Zanette, F. (2004a). Dormancy dynamics of 'imperial gala' apple tree buds during autumn and winter in a region of low chill occurrence. *Revista Brasileira de Fruticultura*, 26: 65-68.

Carvalho, R. I. N., and Zanette, F. (2004b). Dormancy dynamics of two year old buds of 'imperial gala' apple tree in a region of low chill occurrence. *Revista Brasileira de Fruticultura*, 26: 389-391.

Carvalho, R. I. N., and Zanette, F. (2005). Carbohydrate content in buds and stems of apple trees during autumn and winter in a region of low chill occurrence. *Revista Brasileira de Fruticultura*, 26: 202-205.

Carvalho, R. I. N., Biasi, L. A., Zanette, F., Santos, J. M., and Pereira, G. P. (2010). Budburst stages of temperate zone fruit to the biological test of dormancy evaluation. *Revista Academica Ciências Animal*, 8: 93-100.

Carvalho, R. I. N., and Biasi, L. A. (2012). Index for bud dormancy evaluation of temperate climate fruit trees. *Revista Brasileira de Fruticultura*, 34: 936-940.

Champagnat, P. (1983). Bud dormancy, correlation between organs, and morphogenesis in woody plants. *Soviet Plant Physiology*, 30: 458–471.

Citadin, I., Raseira, M. C. B., and Quezada, A. C. (1998). Substrato para conservação de ramos destacado de pessegueiro, Prunus persica L. (Batsh) [Substrate for conservation of outstanding peach tree branches, Prunus persica L. (Batsh)]. *Agropecuária de Clima Temperado*, 1: 55-59.

Citadin, I. (1999). *Necessidade de calor para antese e brotação em pessegueiro Prunus persica (L.) Batsch* [*Heat requirement for anthesis and budburst in peach trees Prunus persica (L.) Batsch*]. MSc diss., Universidade Federal de Pelotas.

Citadin, I., Raseira, M. C. B., Herter, F. G., and Silva, J. B. (2001). Heat requirement for blooming and leafing in peach. *Hortscience,* 36: 305-307.

Citadin, I., Raseira, M. C. B., Herter, F. G., and Silva, J. B., Quezada, A. C., e Silveira, C. A. P. (2002). Meiosis stage as an indicator for peach endodormancy end. *Revista Brasileira de Fruticultura,* 24: 23-28.

Citadin, I., Raseira, M. C. B., Quezada, A. C. and Silva, J. B. (2003). Heritability of heat requirement for blooming and leafing in peach. *Revista Brasileira de Fruticultura*, 25: 119-123.

Citadin, I., Guilliot, A., Bonhomme, M., and Rageau, R. (2009). Walnut enzyme activity related with carbohydrate mobilization during dormancy period. *Revista Brasileira de Fruticultura*, 31: 305-313.

Cook, N. C., Calitz, F. J., Allderman, L. A., Steyn, W. J., and Louw, E. D. (2017). Diverse patterns in dormancy progression of apple buds under variable winter conditions. *Scientia Horticulturae,* 226: 307-15.

Corbesier, L., Prinsen, E., Jacqmard, A., Lejeune, P., Van Onckelen, H., Périlleux, C., and Bernier, G. (2003). Cytokinin levels in leaves, leaf exudate and shoot apical meristem of Arabidopsis thaliana during floral transition. *Journal of Experimental Botany*, 54: 2511-2517.

Crabbe, J. (1970). Influences foliaires sur la croissance de la pousse annuelle du pommier. III Effets de la suppression de jeunes feuilles sur la levee d'inhibition et Ie developpement des bourgeons axillaires [Foliar influences on the growth of the annual growth of the apple tree. III Effects of Young Leaf Removal on the Inhibition Raise and

Development of Axillary Buds]. *Bulletin des Recherches Agronomiques de Gembloux,* 5: 136-151.

Crabbé, J., and Barnola, P. (1996). A new conceptional approach to bud dormancy in woody plants. In *Plant dormancy. physiology, biochemistry and molecular biology*, edited by Lang, G. A., 83-114. Wallingford: CAB International.

Cronjé, P. J. R., Jacobs, G., Sadie, A., and Cook, N. C. (2003). Quantification of the dormancy progression in terminal apple buds. Changes in growth rate and water status. *Advances in Horticultural Science,* 17: 105-110.

Decourteix, M., Alves, G., Bonhomme, M., Peuch, M., Ben B. K., Brunel, N., Guilliot, A., Rageau, R., Améglio, T., Pétel, G., and Sakr, S. (2008). Sucrose (JrSUT1) and Hexoses (JrHT1 and 2) transporters in walnut xylem parenchyma cells: their potential role in early events of growth resumption. *Tree Physiology,* 28: 215–24.

Dennis J., F. G. (2003). Problems in standardizing methods for evaluating the chilling requirements for the breaking of dormancy in buds of woody plants. *Hort Science,* 38: 347-350.

Donoho Junior, Clive W., and Walker, David R. (1957). Effect of gibberellic acid on breaking of rest period in Elberta peach. *Science,* 126: 1178-1179.

Erez, A. (2000). Bud dormancy: Phenomenon, problems and solutions in the tropics and subtropics. In *Temperate fruit crops in warm climates*, edited by Erez, A. 17-48. London: Kluwer Academic Publishers.

Eriksson, M. E., Israelsson, M., Olsson, O., and Moritz, T. (2000). Increased gibberellin biosynthesis in transgenic trees promotes growth, biomass production and xylem fiber length. *Nat Biotechnology,* 18: 784-788.

Fadón, E. Herrero, M., and Rodrigo, J. (2015). Flower development in sweet cherry framed in the BBCH scale. *Scientia Horticulturae,* 192: 141-147.

Fan, S., Bielenberg, D. G., Zhebentyayeva, T. N., Reighard, G. L,, Okie, W. R., Holland, D., and Abbott, A.G. (2009). Mapping quantitative trait lociassociated with chilling requirement, heat requirement and bloomdate in peach (*Prunus persica*). *New Phytologist*, 185: 917–930.

Faust, M., Erez, A., Rowland, L. J., Wang, S. Y., and Norman, H. A. (1997). Bud dormancy in perennial fruit tree: Physiological basis for dormancy induction, maintenance and release. *Hort Science*, 32: 623-629.

Garcia-Bañuelos M. L., Gardea A. A., Winzerling J. J., and Vazquez-Moreno, L. (2009). Characterization of a midwinter-expressed dehydrin (DHN) gene from apple trees (*Malus domestica*). *Plant Molecular Biology Reporter,* 27: 476-487.

Gendraud, M., and Pétel, G. (1990). Modifications in intercellular communications, cellular characteristics and change in morphogenetic potentialities of Jerusalem artichoke tubers (*Helianthus tuberosus* L.). In *Intra and Extracellular Communications in Plants: Reception, Transmission, Storage and Expression of Messages*, edited by Millet, B. and Greppin, H. 171-175. Paris: INRA.

Hartmann, U., Höhmann, S., Nettesheim, K., Wisman, E., Saedler, H., and Huijser, P. (2000). Molecular cloning of SVP: a negative regulator of the floral transition in Arabidopsis. *The Plant Journal,* 21: 351-360.

Hendershott, C. A. (1959). Seasonal fluctuation in quantity of growth substances in resting peach flower buds. *Proceedings of the American Society of Horticultural Science,* 74: 121-129.

Herter, F. G., Rageau, R., Bonhomme, M., e Mauget, J. C. (1992). Determinação do término da dormência e floração para algumas cultivares de macieira: comparação entre métodos biológicos e empírico [Determination of the end of dormancy and flowering for some apple cultivars: comparison between biological and empirical methods]. *Revista Brasileira de Fruticutura,* 14: 77–81.

Herter, F. G., Citadin, I., e Silveira, C. A. P. (2000). Necessidade de calor para a antese em pessegueiro avaliada pelo método de ramos destacados. *Agropecuária Clima Temperado,* 3: 253-259. [Heat requirement for anthesis in peach trees evaluated by the method of highlighted branches. *Agropecuária Clima Temperado,* 3: 253-259].

Ito, A., Sakamoto, D., and Moriguchi, T. (2012). Carbohydrate metabolismo and its possible roles in endodormancy transition in Japanese pear. *Scientia Horticulturae* 144: 187-194.

Ito, A., Sugiura, T., Sakamoto, D., and Moriguchi, T. (2013). Effects of dormancy progression and low-temperature response on changes in the sorbitol concentration in xylem sap of Japanese pear during winter season. *Tree Physiology* 33: 398-408.

Jiménez, S., Lawton-Rauh, A. L., Reighard, G. L., Abbott, A.G. A., and Bielenberg, D. G. (2009). Phylogenetic analysis and molecular evolution of the dormancy associated MADS-box genes from peach. *BMC Plant Biology* 9: 1-12.

Jiménez, S., Reighard, G. L., and Bielenberg D. G. (2010). Gene expression of DAM5 and DAM6 is suppressed by chilling temperatures and inversely correlated with bud break rate. *Plant Molecular Biology* 73:157-167.

Kanayama, Y. (2009). Physiological roles of polyols in horticultural crops. *Journal of the Japanese Society for Horticultural Science*, 78: 158-168, 2009.

Lang, G. A., Early, J. D., Arroyave, N. J., Darnell, R. L., Martin, G. C., and Stutte, G. W. (1985). Toward a reduced, universal terminology. *Horticultural Science*, 20: 809-812.

Lang, G. A. (1987). Dormancy: A new universal terminology. *Hort Science*, 22:817–820.

Lang, G. A., Early, J. D., Martin, G. C., and Darnell, R. L. (1987). Endodormancy, paradormancy, and ecodormancy—physiological terminology and classification for dormancy research. *Hort Science*, 22: 371-377.

Lauri, P. E. (2007). Differentiation and growth traits associated with acrotony in the apple tree (*Malus domestica*, Rosaceae). *American Journal of Botany*, 94: 1273–1281.

Lee, J. H., Yoo. S. J., Park, S. H., Hwang, I., Lee, J. S., and Ahn, J. H. (2007). Role of SVP in the control of flowering time by ambient temperature in Arabidopsis. *Genes Development*, 21: 397-402.

Legave, J. M. (1978). Essai d'interprétation de ne´croses florales avant la floraison chez l'abricotier en relation avec une e´tude des besoins en froid des bourgeons pour la levée de dormance [An interpretation test of pre-flowering floral narcissus in apricot in relation to a study of cold needs of buds for dormancy emergence]. *Annales de l'amélioration des plantes,* 28, 593–607.

Legave, J. M., and Garcia, G., (1982). Chilling requirements for blooming in apricot tree, determined either by a statistical method or by an experimental method and valued according to an exponential conception of temperature action. *Acta Horticulturae*, 121: 103-109.

Legave, J. M., Farrera, I., Almeras, T., Calleja, M. (2008) Selecting models of apple flowering time and understanding how global warming has had an impact on this trait. *The Journal of Horticultural Science and Biotechnology,* 83: 76-84

Lavee, S. (1973). Dormancy and bud break in warm climates: considerations of growth regulator involvement. *Acta Horticuturae,* 34: 225-235.

Lavee, S., and May, P. (1997). Dormancy of grapevine buds – facts and speculation. *Australian Journal of Grape and Wine Research,* 3: 31-46.

Leite, G. B., Bonhomme, M., Putti, G. L., Petel, G., Petri, J. L., and Rageau. R. (2006). Physiological and biochemical evolution of peach leaf buds during dormancy course under two contrasted temperature patterns. *International Journal of Horticultural Science,* 12: 15-19.

Leopold, A. C., and Kriedmann, P. E. (1975). *Plant growth and development*. New York:McGraw Hill.

Li, Z., Reighard, G. L., Abbott, A. G., and Bielenberg, D. G. (2009). Dormancy associated MADS genes from the EVG locus of peach (*Prunus persica* (L.) Batsch) have distinct seasonal and photoperiodic expression patterns. *Journal of Experimental Botany,* 60: 3521-3530.

Luedeling, E., Girvetz, E.H., Semenov, M. A., Brown, P.H. (2011). Climate change affects winter chill for temperate fruit and nut trees. *PLoS One*, 6: 1-13.

Malagi, G. Sachet, M. R., Citadin, I., Herter, F. G., Bonhomme, M., Regnard, J. L., and Legave, J. M. (2015). The comparison of dormancy dynamics in apple trees grown under temperate and mild winter climates imposes a renewal of classical approaches. *Trees,* 29: 1365-1380.

Marafon, A. C., Citadin, I., Amarante, L., Herter, F. G., and Hawerroth, F. J. (2011). Chilling privation during dormancy period and carbohydrate mobilization in Japanese pear trees. *Scientia Agricola,* 68: 462-68.

Marquata, C., Vandamme, M., Gendrauda M., and Pétel, G. (1999). Dormancy in vegetative buds of peach: relation between carbohydrate absorption potentials and carbohydrate concentration in the bud during dormancy and its release. *Scientia Horticulturae,* 79: 151-162.

Mauget, J. C. (1987). Dormance des bourgeons chez les arbres fruitiers de climat tempéré [Dormancy of buds in temperate fruit trees]. In *Le développement des végétaux*, edited by Guyader H., 133-150 Paris: Masson.

Mauget, J. C., and Rageau, R. (1988). Bud dormancy and adaptation of apple tree to mild winter climates. *Acta Horculturae*, 232:101-108.

Maurel, K., Leite, G. B., Bonhomme, M., Guilliot, A., Rageau, R., Petel, G. and Sakr, S. (2004). Trophic control of bud break in peach (Prunus persica) trees: a possible role of hexoses. *Tree Physiology,* 24: 579-588.

Mcpherson, H.G., Snelgar, W. P., Manson, P. J., and Snowball, A. M. (1997). Bud respiration and dormancy of kiwifruit (*Actinidia deliciosa*). *Annals of Botany,* 80:411-418.

Metivier, J. R. (1985). Dormência e Germinação [Dormancy and germination]. In *Fisiologia Vegetal*, editado por Ferri, M. G., 343–392. São Paulo: EPU.

Michaels, S. D., Ditta, G., Gustafson-Brown, C., Pelaz, S., Yanofsky, M. F., and Amasino, R. M. (2003). AGL24 acts as a promoter of flowering in Arabidopsis and is positively regulated by vernalization. *The Plant Journal,* 33: 867-874.

Moing, A., Carbonne, F., Zipperlin, B., Svanella, L., and Gaudillere, J. P. (1997). Phloem loading in peach: Symplastic or apoplastic?. *Physiologia Plantarum,* 101: 489-496.

Morris, D. A., and Arthur, E. D. (1984). Invertase activity in sinks undergoing cell expansion. *Plant Growth Regulation,* 2: 327-337.

Muthalif, M. M., and Rowland, L. J. (1994). Identification of dehydrinlike proteins responsive to chilling in floral buds of blueberry (*Vaccinium,* section *Cyanococcus*). *Plant Physiology,* 104: 1439-1447.

Nir, G., and Shulman, Y. (1986). Changes in the activity of catalase in relation to the dormancy of grape buds. *Plant Physiology,* 81: 1140-1142.

Okie, W. R. and Blackburn, B. (2011). Increase chilling reduce heat requirement for floral budbrake in peach. *Hort Science,* 46 (II): 245-252.

Okuba, H. (2000). Growth cycle and dormancy in plants. In *Dormancy in plants: from whole plant behaviour to cellular control*, edited by Viémont, J. D., Crabbé, J., 1-22. Wallingford: CAB International.

Oukabli, A., and Mahhou, A. (2007). Dormancy in sweet cherry (Prunus avium L.) under Mediterranean climatic conditions. *Biotechnology, Agronomy, Society and Environment,* 11: 133-139.

Paiva, E., and Robitaille, H. A. (1978). Breaking bud rest in detached apple shoots: effects of wounding and ethylene. *Journal of the American Society for Horticultural Science*, 103: 101-104.

Petri, J. L., Palladini, L. A., Shuck, E., Ducroquet, J. H. H. J., and Matos, C. S. (1996). *Dormência e indução da brotação de fruteiras de clima temperado* [*Dormancy and induction of budburst of temperate fruit trees*]. Florianópolis: EPAGRI.

Petri, J. L., and Leite G. B. (2004). Consequences of insufficient winter chilling on apple tree bud-break. *Acta Horticulturae,* 662: 53-60.

Pouget, R. (1963). Recherches physiologiques sur le repos végétatif de la vigne (*Vitis vinifera* L.): la dormance des bourgeons des et le mécanisme de sa disparition [Physiological research on the vegetative rest of the vine (Vitis vinifera L.): the dormancy of the buds and the mechanism of budburst]. *Annales de l'amélioration des plantes,* 13: 1-247.

Putti, G. L. 2001. *Estudo das necessidades de frio e calor para a brotação de cultivares de macieira (Malus domestica, Borck.)* [*Study of the chilling and heat requirements for a budburst of apple trees*]. MSc diss., Universidade Federal de Pelotas.

Rageau, R. (1978). Croissance et débourrement des bourgeons végétatifs de pêncher au cours d'um tes classique de dormance [Growth and budburst of vegetative buds during dormancy]. *Comptes Rendus de l'Académie des Sciences,* 287: 1119-1122.

Richardson, E. A., Seeley, S. D., and Walker, D. R. (1974). A model for estimating the completion of rest for "Redhaven" and "Elberta" peach trees. *Hort Science,* 9: 331-332.

Rinne P. L. H., Kaikuranta P. L. M., Plas L. H. W., and Van Der Schoot, C. (1999). Dehydrins in cold-acclimated apices of birch (*Betula pubescens* Ehrh.): production, localization and potential role in rescuing enzyme function during dehydration. *Planta,* 209: 377-388.

Rodriguez, J., Sherman, W. B., Scorza, R., Wisniewski, M., and Okie, W. R. (1994). Evergreen peach, its inheritance and dormant behavior. *Journal of the American Society for Horticultural Science,* 119: 789-792.

Rogers, H. J. (2017). Leaf senescence In *Encyclopedia of Applied Plant Sciences*, edited by Thomas, B., Murphy, D. J., and Murray, B. G., 308-314. Wellesbourne: Academic Press.

Rohde, A., Prinsen, E., Rycke, R. D., Engler, G., Van Montagu, M., and Boerjan, W. (2002). PtABI3 impinges on the growth and differentiation of embryonic leaves during bud set in poplar. *Plant Cell* 14: 1885–1901.

Rohde, A., and Bhalerao, R. P. (2007). Plant dormancy in the perennial context. *Trends in Plant Science*, 12: 217–223.

Roitsch, T., and Gonzalez, M. C. (2004). Function and regulation of plant invertases: sweet sensations. *Trends Plant Science,* 9: 606-613.

Ruonala, R., Rinne, P. L. H., Baghour, M., Moritz, T., Tuominen, H., and Kangasjärvi, J. (2006). Transitions in the functioning of the shoot apical meristem in birch (*Betula pendula*) involve ethylene. *The Plant Journal,* 46: 628-640.

Salisbury, F. B., Ross, C. W. (1992). *Plant physiology*. Belmont: Wadsworth.

Samish, R. M. (1954). Dormancy in woody plants. *Annales Physiologie Végetale,* 15: 183-204.

Sasaki, R., Yamane, H., Ooka, T., Jotatsu H., Kitamura, Y., Akagi, T., and Tao, R. (2011). Functional and expressional analyses of PmDAM genes associated with endodormancy in Japanese apricot. *Plant Physiology,* 157: 485-497.

Saure, M. C. (1985). Dormancy release in deciduous fruit trees. *Horticultural Reviews,* 7: 239-299.

Sauter, J. J. 1988. Temperature-induced changes in starch and sugars in the stem of *Populus* x *Canadensis* "robusta". *Journal of Plant Physiology,* 132: 608-612.

Sauter, J. J., Wisniewski, M., and Witt, W. (1996). Interrelationships between ultrastructure, sugar levels, and frost hardiness of ray parenchyma cells during frost acclimation and deacclimation in poplar (*Populus* × *canadensis* Moench ‹robusta›) Wood. *Journal of Plant Physiology,* 149: 451-461.

Shaltout, A. D., and Unrath, C. R. (1983). Rest completion prediction model for "Starkrimson Delicious" apples. *Journal of the American Society for Horticultural Science,* 108: 957-961.

Shimizu-Sato, S., and Mori, H. (2001). Control of outgrowth and dormancy in axillary buds. *Plant Physiology,* 127: 1405-1413.

Silveira, C. A. P. (2003). *Avaliação do efeito das horas de frio, épocas de aplicação e concentrações de cianamida hidrogenada e óleo mineral na brotação, floração e frutificação efetiva de pessegueiro em condições de inverno subtropical* [*Evaluation of the effect of chilling hours, times of application and concentrations of hydrogen cyanamide and mineral oil on budburst, flowering and effective fruiting of peach in subtropical winter conditions*]. PhD diss., Universidade Federal de Pelotas.

Stafstrom, J. P. (2000). Regulation of growth and dormancy in pea axillary buds." In *Dormancy in plants: From whole plant behaviour to cellular control*, edited by Viémont, J. D., and Crabbé, J., 331-346. Wallingford: CAB International.

Stirnberg, P., Van De Sande, K., and Leyser, O. H. M. (2002). MAX1 and MAX2 control shoot branching in Arabidopsis. *Development,* 129: 1131-1141.

Szyjewicz, E., Rosner, N., and Kliewer, W. M. (1984). Ethephon ((2-Chloroethyl) phosphonic Acid, Ethrel, CEPA) in viticulture: a review. *American Journal of Enology and Viticulture,* 35: 117-123.

Tabuenca, M. C. (1964). Necesidades de frio invernal de variedades de albaricoquero, melocotonero y peral [Chilling requirement of apricot, peach and pear varieties]. *Anual Estación Experimental de Aula Dei,* 7: 113-132.

Tabuenca, M. C. (1967). Necesidades de frio invernal de variedades de ciruelo [Chilling requirement of plum varieties]. *Anual Estación Experimental de Aula Dei,* 8: 383-391.

Tamura, F., Tanabe, K., Itai, A., and Tanaka, H. (1998). Protein changes in the flower buds of japanese pear during breaking of dormancy by chiling or high-temperature treatment. *Journal of the American Society for Horticultural Science,* 123: 532-536.

Tanino, K. K., Kalcsits, L., Silim, S., Kendall, E., and Gray, G. R. (2010). Temperature-driven plasticity in growth cessation and dormancy development in deciduous woody plants: a working hypothesis suggesting how molecular and cellular function is affected by temperature during dormancy induction. *Plant Molecular Biology*, 73: 49-65.

Tort, M., Gendraud, M., and Courduroux, J. C. (1985). Mechanisms of storage in dormant tubers: correlative aspects, biochemical and ultrastructural approaches. *Physiologie végétale,* 23: 289-299.

Tromp, J., Webster, A. D., and Wertheim, S. J. (2005). Fundamentals of temperature zone tree fruits production. *Backhuys Publishers,* 25: 1571-1572.

Vegis, B. A. (1964). Dormancy in Higher Plants. *Annual Review of Plant Physiology,* 15: 185-224.

Viti, R., and Monteleone, P., (1993). High temperature influence on the presence of flower bud anomalies in two apricots varieties characterized by different productivity. *Acta Horticulturae,* 384: 283–289.

Wagner Júnior, A., Bruckner, C.H., Salomão, L.C.C., Pimentel, L.D., Silva, J. O. C., and Santos, C. E. M. Evaluation of chilling requirement in peach through grafted twigs. *Revista Brasileira de Fruticultura*, 31: 1054-1059, 2009.

Wang, S. Y., and Faust, M. (1990). Changes of membrane lipids in apple buds during dormancy and budbreak. *Journal of the American Society for Horticultural Science*, 115, 803–808.

Weinberger, J. H. (1967). Studies on flower bud drop in peaches. *Proceedings of the American Society for Horticultural Science,* 91:78-83.

Weinberger, J. H. (1969). The stimulation of dormant peach bud by a cytokinin. *Hort Science,* 4: 125-126.

Wisniewski, M., Webb, R., Balsamo, R., Close, T. J., Yu, X. M., and Griffith, M. (1999). Purification, immunolocalization, cryoprotective, and antifreeze activity of PCA60: A dehydrin from peach (*Prunus persica*). *Physiologia Plantarum,* 105: 600-608.

Wisniewsk, M., Bassett, C., and Arora, R. (2004). Distribution and partial characterization of seasonally expressed proteins in different aged shoots and roots of 'Loring' peach (*Prunus persica*). *Tree Physiology*, 24: 339-345.

Yamane, H., Kashiwa, Y., Ooka, T., Tao, R., and Yonemori, K. (2008). Suppression subtractive hybridization and differential screening reveals endodormancy-associated expression of an SVP/AGL24-type MADS-box gene in lateral vegetative buds of Japanese apricot. *Journal of the American Society for Horticultural Science,* 133: 708-716.

Yamane, H., Ooka, T., Jotatsu, H., Hosaka, Y., Sasaki R., and Tao, R. (2011a). Expressional regulation of PpDAM5 and PpDAM6, peach (*Prunus persica*) dormancy-associated MADS-box genes, by low temperature and dormancy-breaking reagent treatment. *Journal of Experimental Botany*, 62: 3481–3488.

Yamane, H., Ooka, T., Jotatsu, H., Sasaki, R., and Tao. R. (2011b). Expression analysis of PpDAM5 and PpDAM6 during flower bud development in peach (*Prunus persica*). *Scientia Horticulturae*. 129: 844–848.

Zhong, W., Gao, Z., Zhuang, W., Shi, T., Zhang, Z., and Ni, Z. (2013). Genome-wide expression profiles of seasonal bud dormancy at four critical stages in Japanese apricot. *Plant Molecular Biology*, 83: 247-264.

About the Editor

Renato Vasconcelos Botelho is an Agronomist, with Master and Doctor degrees in Agronomy (Horticulture) by the Faculty of Agronomic Sciences at the State University of São Paulo (Unesp), Campus Botucatu, Brazil. He has developed postdoctoral projects in the section of Fruit Tree Crops of the Agronomic Institute of Campinas (2003) and in the School of Agrarian and Veterinary of the University of Bologna, Italy (2013). He is an associate professor at the State University of MidWestern of Paraná, in Guarapuava, Paraná, Brazil; where he has been a faculty member since 2003. He is a faculty member of the Postgraduate Program in Agronomy of the Ponta Grossa State University (UEPG), Brazil, since 2011. He develops research in the area of fruit trees growing in the organic system, working mainly in the following topics: plant dormancy, viticulture, deciduous fruit trees, alternative control of fruit diseases, nutrition, soil fertility, fertilization, native fruits, post-harvest physiology and technology.

INDEX

A

B

C

R

S

T

V

W